Macromolecular Symposia | 257

Polyolefin Characterization

Selected Contributions
from the conference:
The First International Conference on
Polyolefin Characterization (ICPC)
Houston, TX (USA), October 16–18, 2006

Symposium Editor:
João B. P. Soares
(University of Waterloo, Canada)

Elias, H.-G.

Macromolecules
Volume 3: Physical Structures and Properties

approx. 693 pages with approx.
360 figures and approx. 100 tables 2007,
Hardcover
ISBN: 978-3-527-31174-3

Elias, H.-G.

Macromolecules
Volume 4: Applications of Polymers

approx. 660 pages with approx.
270 figures and approx. 170 tables 2008,
Hardcover
ISBN: 978-3-527-31175-0

Meyer, T., Keurentjes, J. (eds.)

Handbook of Polymer Reaction Engineering

1137 pages in 2 volumes with 373 figures
and 96 tables 2005, Hardcover
ISBN: 978-3-527-31014-2

Odian, G.

Principles of Polymerization

832 pages
2004, Hardcover
ISBN: 978-0-471-27400-1

M. Buback, A. M. V. Herk (Eds.)

Radical Polymerization: Kinetics and Mechanism

Macromolecular Symposia 248
ISBN: 978-3-527-32056-1

R. A. Hutchinson (Ed.)

Polymer Reaction Engineering VI

Macromolecular Symposia 243
ISSN: 1022-1360

Macromolecular Symposia | Vol. 257

Polyolefin Characterization

Selected Contributions
from the conference:
The First International Conference on
Polyolefin Characterization (ICPC)
Houston, TX (USA), October 16–18, 2006

Symposium Editor:
João B. P. Soares
(University of Waterloo, Canada)

ISBN 3-527-32192-6

WILEY-VCH

Full text and further information: www.ms-journal.de

Macromolecular Symposia
is published 14 times a year

Annual subscription rates 2008
Macromolecular Full Package
(All seven *Macromolecular Journals*; 101 issues in total):

Europe	Euro	8,999	9,899
Switzerland	Sfr	14,995	16,495
All other areas	US$	11,895	13,085
		print only **or** electronic only	print **and** electronic

Postage and handling charges included.
All Wiley-VCH prices are exclusive of VAT.
Prices are subject to change.
Individual subscriptions, single issues and back copies are available.
Please ask for details at: service@wiley-vch.de

Orders may be placed through your bookseller or directly at the publishers:
WILEY-VCH Verlag GmbH & Co. KGaA,
P.O. Box 10 11 61, 69451 Weinheim, Germany,
Tel. +49 (0) 62 01/6 06-400,
Fax +49 (0) 62 01/60 61 84,
E-mail: service@wiley-vch.de

Copyright Permission:
Fax: +49 (0) 62 01/6 06-332,
E-mail: rights@wiley-vch.de

For USA and Canada: Macromolecular Symposia (ISSN 1022-1360) is published with 14 volumes per year by WILEY-VCH Verlag GmbH & Co. KGaA, Boschstr. 12, 69451 Weinheim, Germany. Air freight and mailing in the USA by Publications Expediting Inc., 200 Meacham Ave., Elmont, NY 11003, USA. Application to mail at Periodicals Postage rate is pending at Jamaica, NY 11431, USA. POSTMASTER please send address changes to: Macromolecular Symposia, c/o Wiley-VCH, 111 River Street, Hoboken, NJ 07030, USA.

Printed on acid-free paper.
Typesetting: Thomson Digital (India) Ltd., India
Printing: Strauss Offsetdruck, Mörlenbach
Binding: Litges & Dopf, Heppenheim

Macromolecular Symposia: Vol. 257

Articles published on the web will appear several weeks before the print edition. They are available through:

WILEY InterScience®

www.interscience.wiley.com

Cover: Tridimensional surface plot of a multiple-reactor polyethylene sample measured with TREF-GPC cross-fractionation (lower left corner) and its two-dimensional projections showing the MWD and average TREF elution temperature as a function of molecular weight (upper left corner) and TREF profile and log M_w as a function of elution temperature (lower right corner).

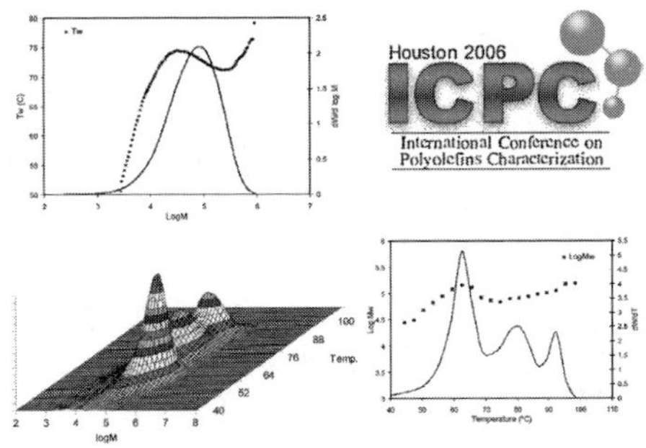

Polyolefin Characterization

Houston, TX (USA), October 16-18, 2006

The final properties and applications of polyolefins are determined by their microstructures. This statement can be made for any polymer, but it is even more relevant for polyolefins since they are composed of such simple building blocks. Therefore, it is no surprise that the study of polyolefin microstructure has always been an essential part of catalyst research, process troubleshooting and optimization, and product development. In addition to conventional Ziegler-Natta polyolefins, which already have rather intricate microstructures, there is a rapidly growing interest in producing polyolefins with increasingly more complex molecular architectures, using combinations of single and multiple-site catalysts and/or polymerization processes with two or more reactors operated in series at different conditions. Recent examples are linear-block and branch-block copolymers with elastomeric properties, copolymers with long chain branches made by one or more single-site catalysts, and bimodal resins made in tandem or multizone reactor processes. The increasing sophistication of catalyst systems and polymerization processes can only be fully realized in practice if efficient and easy to use polyolefin characterization techniques are available. The First International Conference on Polyolefin Characterization (ICPC) was organized to fill this important industrial and academic need, providing a discussion forum on the characterization and fractionation techniques of polyolefins.

The first ICPC took place in Houston, TX, from October 16 to 18, 2006. One hundred and seven participants from 18 different countries attended the conference: 65 from the industry, 24 from academia, and 18 from vendor companies. The strong participation from the polymer manufacturing industry from North America, Europe and Asia shows the industrial relevance and need of such a conference. In addition, a one-day course on polyolefin characterization techniques was offered before the beginning of the conference for those participants interested in an update on the principles of gel permeation chromatography (GPC), temperature rising elution fractionation (TREF), and crystallization analysis fractionation (CRYSTAF).

The oral presentations given during the 1st ICPC were divided according to main topic areas into Separation and Fractionation, High Throughput, Thermal and Crystallinity Analysis, Spectroscopy, and Rheology. In addition to the oral presentations, 29 posters were displayed.

We intend to continue organizing the ICPC biannually, alternating between North American and European locations. The 2nd ICPC will take place in September 2008 (exact dates are still to be defined) in Valencia, Spain. Readers interested in being included in the conference's mailing list are welcome to send us their requests by e-mail to *raquel.ubeda@icpc-conference. org*

João B. P. Soares

ICPC organizing committee members: Benjamin Monrabal, Raquel Ubeda, Colin Li Pi Shan, and João B. P. Soares.

An Overview of Important Microstructural Distributions for Polyolefin Analysis

João B.P. Soares

Summary: Polyolefins with complex microstructures are becoming increasingly common in academic and industrial applications. Polyolefin analytical techniques are evolving to provide a more detailed picture of these microstructures, with the development and improvement of hyphenated-techniques and cross-fractionation methods. These modern analytical techniques provide a wealth of information on polyolefin microstructure and, despite being extremely useful, they can also be hard to interpret without the help of mathematical models that link polymerization kinetics to chain microstructure and polymer characterization results. In this paper we review some of the most important distributions for polyolefin microstructure and derive a few new expressions that help understand the results obtained with several polyolefin characterization techniques.

Keywords: polyethylene; polymer characterization; polymer fractionation; polymer microstructure; polyolefins

Introduction

The remarkable versatility of polyolefins come from the fact that ethylene, propylene and α-olefins can be copolymerized to create polymer chains with microstructures that lead to very different physical properties.

Polyolefin properties are ultimately defined by the way the monomers are connected to form linear and branched polymer chains with different degrees of regularity. It is, therefore, very important to characterize the microstructure of polyolefins and to quantify this microstructure using fundamental models.

In this short overview, we will present some important equations that describe polyolefin microstructure and discuss some modeling principles that can be used to help understand the results obtained with several polyolefin characterization techniques.

Department of Chemical Engineering, University of Waterloo, Waterloo, Ontario, Canada N2L 3G1
E-mail: jsoares@uwaterloo.ca

Distribution of Chain Length, Chemical Composition, and Long Chain Branching

The most general distribution for the microstructure of polyolefins made with coordination catalysts is given by the equation:[1,2]

$$w(r,F,i) = \frac{1}{(2i+1)!} r^{2i+1}\tau^{2i+2} \exp(-r\tau)\sqrt{\frac{r}{2\pi\beta}}\exp\left[-\frac{r(F-\overline{F})^2}{2\beta}\right] \quad (1)$$

In Equation (1), $w(r,F,i)$ is the height of the weight distribution for chains of length r, comonomer fraction F, and i long chain branches (LCB) per chain. This equation has only two parameters, β and τ, defined as:

$$\beta = \overline{F}(1-\overline{F})\sqrt{1-4\overline{F}(1-\overline{F})(1-r_1r_2)} \quad (2)$$

$$\tau = \frac{\textit{rate of transfer} + \textit{rate of LCB formation}}{\textit{rate of propagation}} \quad (3)$$

Finally, $\overline{F}$ is the average fraction of comonomer in the copolymer (as calculated from Mayo-Lewis equation, for instance) and r_1 and r_2 are the comonomer reactivity ratios.

This equation was derived based on the mechanism widely accepted for olefin polymerization with coordination catalysts where chains can propagate by monomer insertion, terminate through several transfer mechanism, and LCBs are formed by the incorporation of vinyl-terminated polymer chains, commonly called macromonomers. No other assumption were needed for model development.[1,2]

It is interesting to point out that Equation (1) becomes Stockmayer distribution for linear chains, that is, for $i=0$:[3]

$$w(r,F) = r\tau^2 \exp(-r\tau)\sqrt{\frac{r}{2\pi\beta}}\exp\left[-\frac{r(F-\overline{F})^2}{2\beta}\right] \quad (4)$$

In addition, if we integrate Equation (4) over all comonomer compositions we obtain Flory's most probable chain length distribution (CLD):[4,5]

$$w(r) = \int_{-\infty}^{\infty} w(r,F)\mathrm{d}(F-\overline{F}) = r\tau^2\exp(-r\tau) \quad (5)$$

For linear chains, the parameter τ is the reciprocal of the number average chain length, r_n:

$$\tau = \frac{\textit{rate of transfer}}{\textit{rate of propagation}} = \frac{1}{r_n} \quad (6)$$

Therefore, in the same way that Stockmayer's distribution is the extension of Flory's distribution to binary copolymers, Equation (1) is the extension of Stockmayer's distribution to non-linear copolymers. We will now start applying Equation (1), (4), and (5) to several common polyolefin characterization techniques.

Molecular Weight Distribution of Linear Chains

Molecular weight distributions (MWD) of polyolefins made with single site catalysts follow Flory's most probable distribution, Equation (5). MWDs are usually measured with high-temperature gel permeation chromatography (GPC) and expressed in log scale. Before we can use Equation (5) to describe the experimental MWD of polyolefins, we need to apply two simple mathematical transformations. First, we need to change the CLD into a MWD using the relation,

$$w(MW)\mathrm{d}MW = w(r)\mathrm{d}r \quad (7)$$

where MW is the polymer molecular weight. Since $\mathrm{d}MW/\mathrm{d}r$ equals the molar mass of the repeating unit in the polymer chain (mw), Equation (5) becomes,

$$w(MW) = MW\hat{\tau}^2\exp(-MW\hat{\tau}) \quad (8)$$

where,

$$\hat{\tau} = \frac{\tau}{mw} = \frac{1}{r_n \cdot mw} = \frac{1}{M_n} \quad (9)$$

and M_n is the number average molecular weight of the polymer.

Equation (8) must now be rendered in log scale through the transformation:

$$w(\log MW)\mathrm{d}\log MW = w(MW)\mathrm{d}MW \quad (10)$$

Consequently:

$$w(\log MW) = 2.3026 \times MW^2\hat{\tau}^2\exp(-MW\hat{\tau}) \quad (11)$$

Figure 1 shows that the MWD of a polyethylene sample made with two metallocene catalysts supported on the same silica carrier is well represented by the superposition of two Flory's distributions. Since we have two single-site catalysts in Figure 1, the MWD of the combined polymer, $W(\log MW)$ is described by the weighed sum of two Flory's distributions,

$$W(\log MW) = m_{Zr}w(\log MW)_{Zr} + (1-m_{Zr})w(\log MW)_{Hf} \quad (12)$$

where m_{Zr} is the mass fraction of polyethylene produced by the zirconium catalyst.

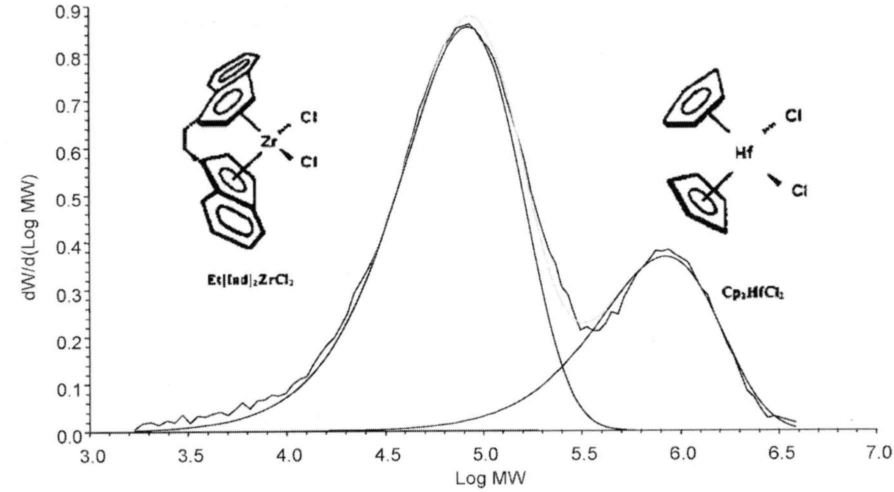

Figure 1.
Comparison of the GPC-measured MWD of a polyethylene sample made with two metallocenes supported on silica and model prediction using Flory's distribution. The peaks for polymer made with both metallocenes are described with Equation (8).[6]

This modeling concept is commonly extended to catalysts that have more than one site type, such as heterogeneous Ziegler-Natta and Phillips catalysts.[7] Figure 2 shows an example of such a MWD representation for the case of a polyethylene resin made with a heterogeneous Ziegler-Natta catalyst.

Therefore, we can generalize Equation (12) for the case of a catalyst with n

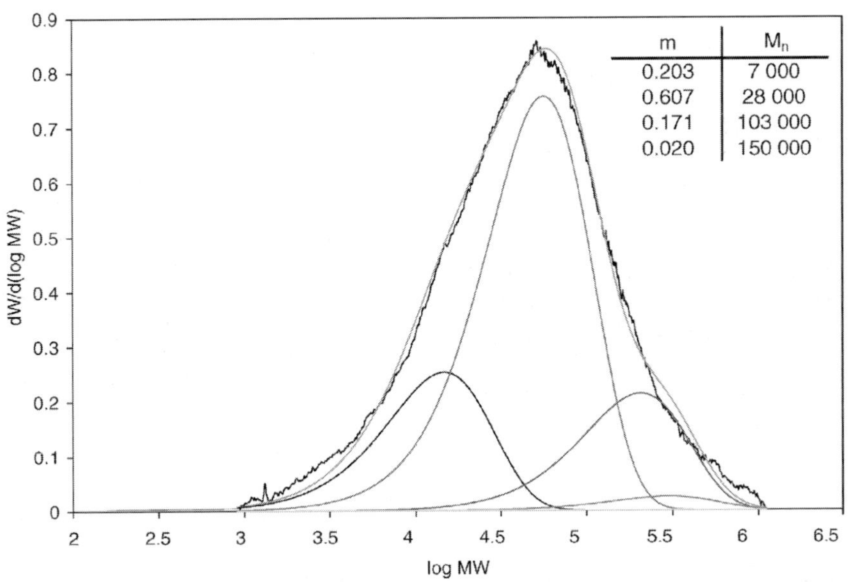

m	M_n
0.203	7 000
0.607	28 000
0.171	103 000
0.020	150 000

Figure 2.
MWD of a polyethylene sample made with a heterogeneous Ziegler-Natta catalyst. The MWD is represented as a superposition of four Flory's distributions, having the mass fractions (m) and number average molecular weights (M_n) indicated in the table.

different site types:

$$W(\log MW) = \sum_{j=1}^{n} m_j w_j(\log MW)$$
$$= 2.3056$$
$$\times MW^2 \sum_{j=1}^{n} m_j \hat{\tau}_j^2 \exp(-MW\hat{\tau}_j) \qquad (13)$$

Equation (13) is a statement of our first modeling principle:

Principle 1: The microstructural distribution of a polymer made with a multiple-site catalyst can be represented as a weighted superposition of distributions for single-site catalysts.

This principle must be used with care: we must keep in mind that it only provides a convenient way to represent microstructural distributions of polymers made with multiple site catalysts.

Hyphenated Techniques: GPC-IR

The use of a infrared detector (IR) with GPC is becoming increasingly popular for polyolefin characterization. This relatively simple combination permits the detection of the average chemical composition (generally reported as molar fraction of α-olefin or short chain branching frequency) as a function of molecular weight. Figure 3 shows the GPC-IR plot for a linear low density polyethylene (LLDPE) resin. We immediately notice the fingerprint mark of a heterogeneous Ziegler-Natta catalyst: as the molecular weight increases, the fraction of 1-butene in the sample decreases.

It is possible to use Modeling Principle 1 to interpret this profile. Figure 4 shows that the MWD can be represented as a superposition of five Flory's distributions. If we assume that each distribution is associated with an active site type that produces LLDPE with a distinct average molar fraction of 1-butene (F_j), we can say that the overall 1-butene fraction measured by the IR detector in a given molecular weight (ΔMW) interval is,

$$F(\Delta MW) = \sum_{j=1}^{5} \Delta w_j(\Delta MW) F_j \qquad (14)$$

where Δw_j is the mass fraction of polymer made on site type j eluting from the GPC column set in the interval ΔMW. The mass fractions Δw_j are obtained from the integration of the Flory distribution associated with each site type:

$$\Delta w_j(MW) = m_j \int_{MW}^{MW+\Delta MW} w_j(MW) \mathrm{d}MW$$
$$= m_j \left\{ \begin{array}{l} (1 + MW\hat{\tau}_j)\exp(-MW\hat{\tau}_j) \\ -[1 + (MW + \Delta MW)\hat{\tau}_j] \\ \exp[-(MW + \Delta MW)\hat{\tau}_j] \end{array} \right\} \qquad (15)$$

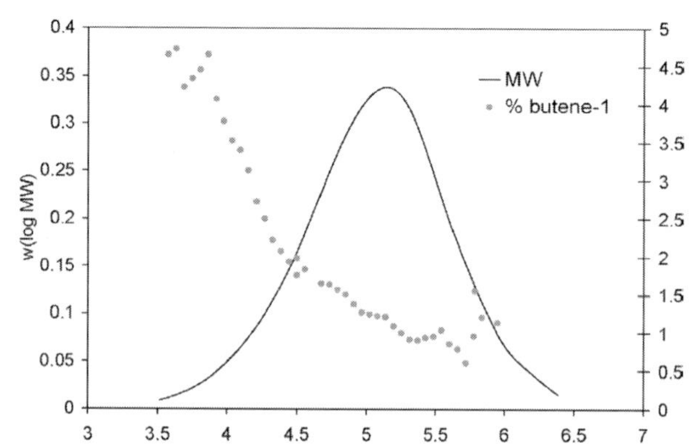

Figure 3.
GPC-IR plot of a LLDPE resin.

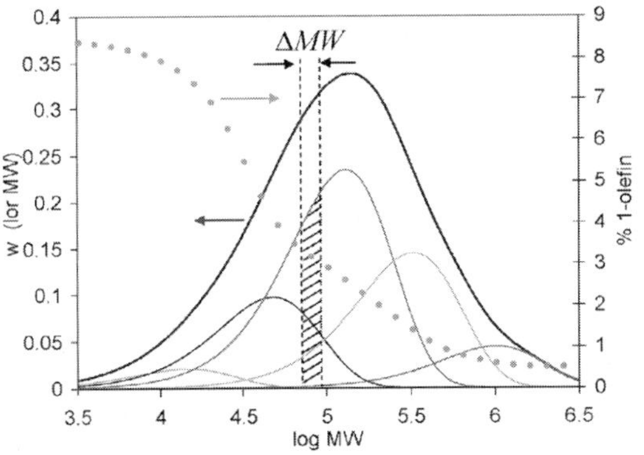

m	M_n	F
0.0382	7 000	0.09
0.18	24 000	0.04
0.433	65 000	0.03
0.2674	164 000	0.01
0.0815	510 000	0.005

Figure 4.
GPC-IR representation using five site types. The table lists the mass fractions (*m*), number average molecular weights (M_n), and 1-olefin fraction (*F*) of polymer made on each site type.

Notice that for the low molecular weight region, one may have to account for the effect of methyl end groups on the experimental IR data.

This simple representation of GPC-IR profiles permit a systematic interpretation of results observed in several academic and industrial polyolefin analytical laboratories.

Chemical Composition Distribution of Linear Chains

The bivariate distribution of chain length and chemical composition of linear polyolefins is given by Stockmayer's distribution, Equation (4). A short description of its main features is useful to clarify several properties of binary copolymers such as LLDPE and propylene/ethylene copolymers.

Figure 5 shows Stockmayer's distributions for four model single-site polyolefins with the same reactivity ratio product ($r_1r_2 = 1$, random copolymers) and average ethylene fraction ($\overline{F} = 0.8$), but with different average chain lengths. Notice that, as the number average chain lengths of the samples increase, their distributions become narrower on the chemical composition dimension. This trend is also observed for each sample individually: shorter chains have a broader chemical composition distribution (CCD) than longer chains. This is a well known effect, caused by the statistical averaging of the chemical compositions per chain as the chains get longer. Samples with infinite length would all have comonomer fractions exactly equal to the average comonomer fraction of the polymer.

The other important property of Stockmayer's distribution is shown in Figure 6: the CCD component broadens steadily when the reactivity ratio product increases, that is, as the copolymer passes from alternating to random and, finally, to block comonomer sequences. This is also an intuitive concept, since all chains of a perfectly alternating copolymer have the same composition ($F = 0.5$), while a tendency to form long blocks of one of the comonomer will necessary increase intermolecular heterogeneity.

We can apply our Modeling Principle 1 to Stockmayer's distribution to describe the bivariate distribution of chain length (or molecular weight) and chemical composition of polyolefins made with multiple site catalysts. In this case, the following generic expression applies,

$$W(r, F) = \sum_{j=1}^{n} m_j w_j(r, F) \tag{16}$$

where $w_j(r,F)$ for each site is given by Equation (4). It should be clear that this equation can be transformed into a molecular weight distribution and expressed in either linear or log scale, using the transformations demonstrated above for Flory's distribution.

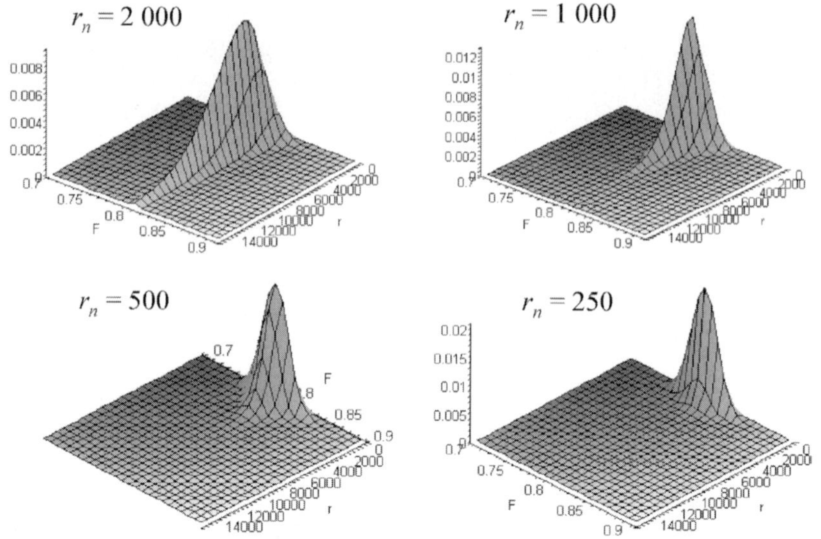

Figure 5.
CLD and CCD of four model single-site polyolefin with different number average chain lengths, r_n. Model parameters: $\bar{F} = 0.8$, $r_1r_2 = 1$, and $\tau = 1/r_n$.

Figure 7 illustrates two bivariate distributions for LLDPE resins. The experimental distribution on the left side was measured using Polymer *Ch*ar CFC 300, a cross fractionation instrument that combines fractionation by temperature rising elution fractionation (TREF) and GPC,[8] while the distribution on the right side of the figure is a model representation using Equation (16) for a three site-type catalyst.

Very few polyolefin characterization laboratories have cross-fractionation instruments available, but most have either a TREF and/or a crystallization analysis fractionation (Crystaf) unit. We can obtain the CCD component of Stockmayer's distribution with the integration:

$$w(F) = \int_0^\infty w(r, F)\mathrm{d}r = \frac{3}{4\sqrt{2\beta\tau}\left[1 + \frac{(F-\bar{F})^2}{2\beta\tau}\right]^{5/2}} \tag{17}$$

We will use Equation (17) to help us define our second modeling principle:

Principle 2: Individual microstructural distributions can be obtained from the integration of multivariate distributions.

Notice that we had already used Modeling Principle 2 to isolate the CLD component, Equation (4), from Stockmayer's distribution.

Figure 8 shows how the breadth of the CCD depends on the product of the parameters β and τ. The distribution broadens as the polymer chains become shorter (increasing τ) or the copolymer chains become blockier (increasing β). These trends had already been described in Figure 5 and 6, and appear now as part of a lumped parameter given by the product $\beta\tau$. Figure 8 captures the essence of olefin copolymer microstructure in a very elegant plot.

Similarly to the procedure we adopted to describe the MWD of polyolefins made with multiple-site catalysts, the CCD of polyolefins made with these catalysts can be represented as a weighted superposition of single-site CCDs:

$$W(F) = \sum_{j=1}^{n} m_j w_j(F) \tag{18}$$

The MWD and CCD (measured as TREF elution profiles) of an ethylene/1-butene copolymer made with a heterogeneous Ziegler-Natta catalyst is shown in

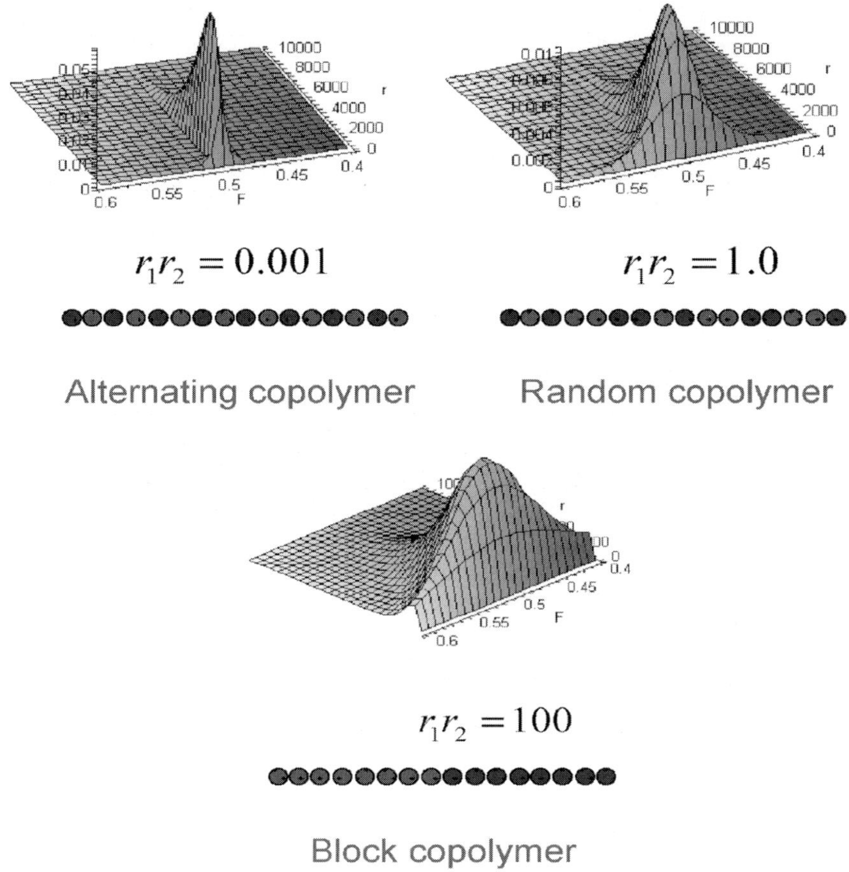

Figure 6.
CLD and CCD of three model single-site polyolefin with different reactivity ratio products, r_1r_2. Model parameters: $\bar{F} = 0.5$, $\tau = 0.001$.

Figure 9. Both profiles can be well represented with four site types. It should be noticed, however, that TREF and Crystaf profiles are related to, but are not in fact the CCD described by Equation (17). Therefore, the mathematical treatment illustrated in Figure 9 is only a first over simplified approximation. The development of fundamental models for TREF and Crystaf is a hard subject that is beyond the scope of this short review.[9,10]

Hyphenated Techniques: TREF-LS

When a light-scattering detector is added to TREF, it is possible to measure the weight average molecular weight (M_w) of polyolefin as a function of elution temperature or comonomer fraction. This technique provides a complementary analysis to the other hyphenated technique, GPC-IR, described above. Similar information can also obtained by projecting the CFC results onto the TREF elution temperature plane and computing the M_w for each elution temperature.

Figure 10 shows one analytical result for an ethylene/α-olefin copolymer with rather complex microstructure. We notice that there is a strong correlation between the elution temperature (or comonomer fraction) and the weight average molecular weigh of the polymer. The trends for the

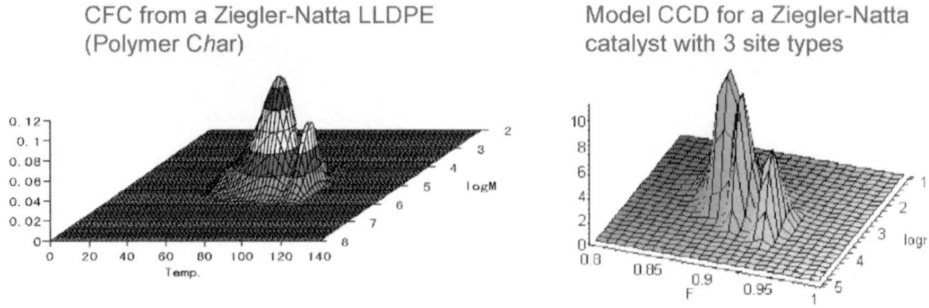

Figure 7.
Bivariate distributions for experimental (left) and model (right) LLDPE resins. The temperature axis on the experimental distribution can be converted into a comonomer fraction.

low crystallinity peak (indicated in the oval section in Figure 10) are particularly interesting: M_w passes through a maximum as we go across the peak.

Once again, the fundamental microstructural models we have been working with in this article can help us understand this trend. The weight average chain length (r_w) of a polymer as a function of its composition can be calculated from Stockmayer's distribution as:

$$r_w(F) = \int_0^\infty r w(r,F)\mathrm{d}r = \frac{15\beta^3\tau^2}{[2\beta\tau + (F-\overline{F})^2]^{7/2}} \quad (19)$$

Similarly, the number average chain length (r_n) is calculated as:

$$r_n(F) = \int_0^\infty r f(r,F)\mathrm{d}r = \frac{3\beta^2\tau}{[2\beta\tau + (F-\overline{F})^2]^{5/2}} \quad (20)$$

where $f(r,F)$ is Stockmayer's *molar* distribution:

$$f(r,F) = \frac{w(r,F)}{r\tau} \quad (21)$$

Finally, the polydispersity index as a function of chemical composition is given by:

$$PDI(F) = \frac{r_w(F)}{r_n(F)} = \frac{5\beta\tau}{2\beta\tau + (F-\overline{F})^2} \quad (22)$$

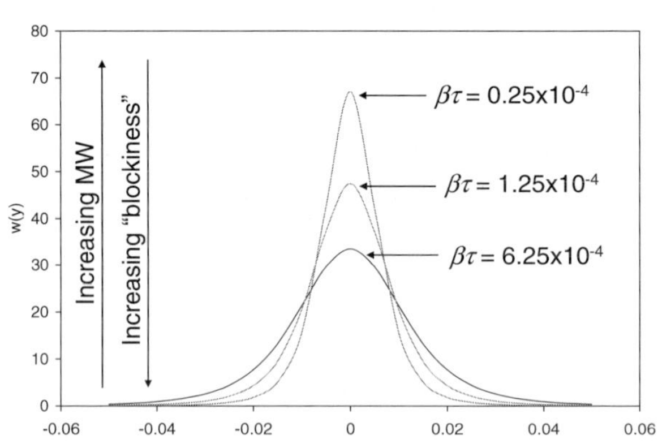

Figure 8.
CCD component of Stockmayer's distribution. The lumped parameter $\beta\tau$ determines the breadth of the CCD.

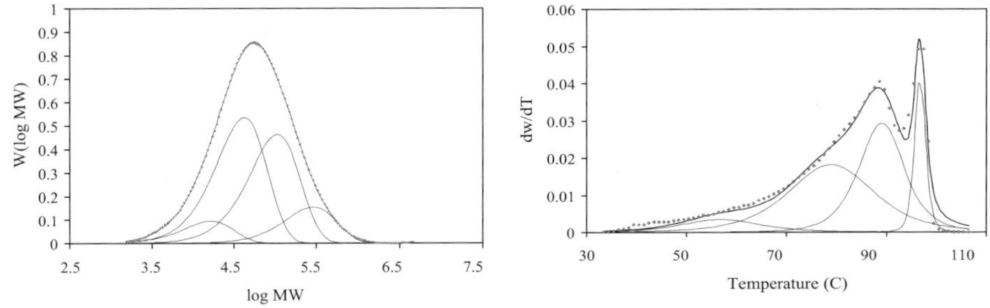

Figure 9.
MWD and CCD of a heterogeneous Ziegler-Natta ethylene/1-butene copolymer represented with four single-site populations. Model parameters: $m_1 = 0.074$, $m_2 = 0.428$, $m_3 = 0.372$, $m_4 = 0.126$; $M_{n1} = 8\,300$, $M_{n2} = 21\,000$, $M_{n3} = 53\,000$, $M_{n4} = 150\,000$; $T_1 = 57\,^{\circ}C$, $T_2 = 79\,^{\circ}C$, $T_3 = 89\,^{\circ}C$, $T_4 = 97\,^{\circ}C$. (The TREF elution temperature can be translated into 1-butene fraction with a calibration curve that accounts for the effect of 1-butene molar fraction and blockiness.)[11]

Equation (19) and (22) are shown in Figure 11, together with a CCD of a model binary copolymer. Notice that the trends predicted for r_w are the same as the ones measured for the M_w of the lowest crystallinity peak in Figure 10. This result confirms one of our previous observations with Stockmayer's distribution that the comonomer fraction in the longer chains approximates more closely the average comonomer fraction in the whole copolymer. Since the peak position in TREF is associated with the average comonomer fraction in the copolymer, we should expect that the molecular weight should increase as we get closer to this peak. Interestingly, the PDI also passes through a maximum value of $PDI = 2.5$ when $F = \overline{F}$ at the peak position for a polymer made with a single site catalyst. To our knowledge, this theoretical prediction has not been validated experimentally yet.

Our third and last modeling principle can now be defined:

Principle 3: Average properties can be obtained from a bivariate distribution by integrating over one of its dimensions.

The TREF-M_w profile for the multisite catalyst shown in Figure 10 can now be described as a superposition of CCD curves

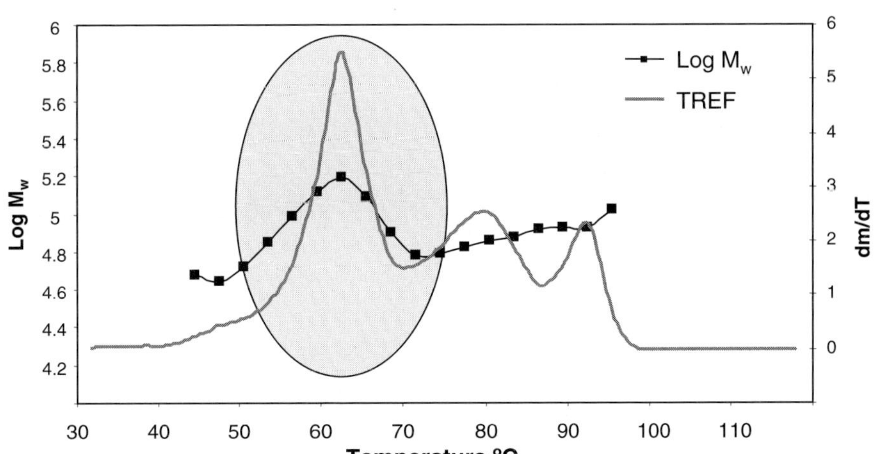

Figure 10.
TREF – M_w plot of an ethylene/α-olefin copolymer with complex microstructure (Polymer Char).

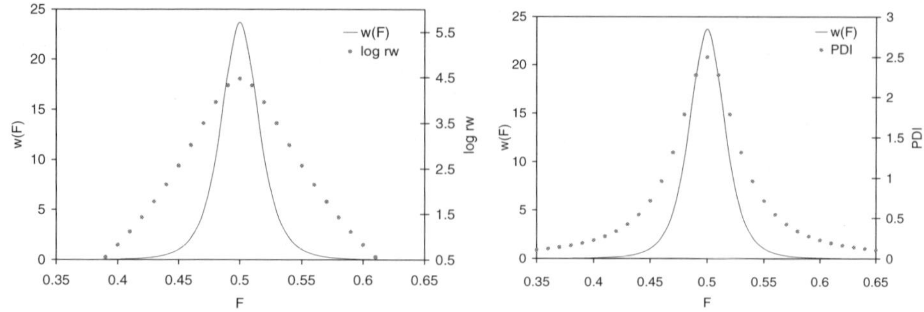

Figure 11.
Weight average chain length (r_w) and polydispersity index (*PDI*) of a binary model copolymer as a function of comonomer composition. Model parameters: $r_n = 500$, $\bar{F} = 0.5$, $r_1 r_2 = 1$.

and average weight average chain lengths (R_w) with the equation,

$$R_w(F) = \sum_{j=1}^{n} \Delta w_j(F) m_j r_{w_j}(F) \tag{23}$$

where $\Delta w_j(F)$ is the fraction of polymer with comonomer composition F made on site type j, calculated as:

$$\Delta w_j(F) = \int_{F}^{F+\Delta F} w_j(F) \mathrm{d}F \tag{24}$$

Figure 12 shows a model plot for three site types. These results are only qualitative; they represent trends, not actual values, since the representation of TREF profiles directly from Stockmayer's distributions is not accurate enough for quantitative calculations.

Long Chain Branching Distribution

Figure 13 shows the predicted distributions of chain length and comonomer composi-

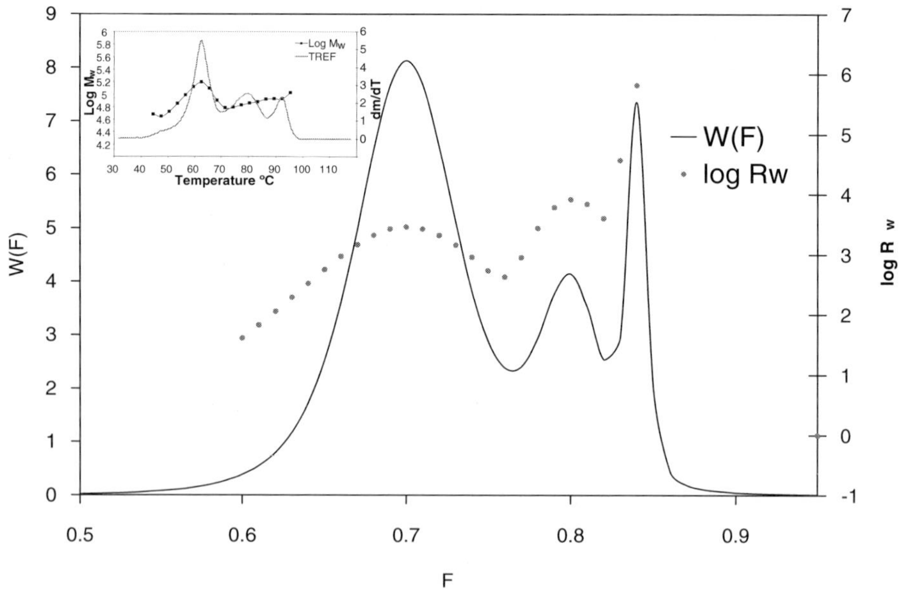

Figure 12.
CCD-R_w profile for a model three-site copolymer. The theoretical plot was adjusted to resemble the experimental plot (insert) but this representation is only qualitative.

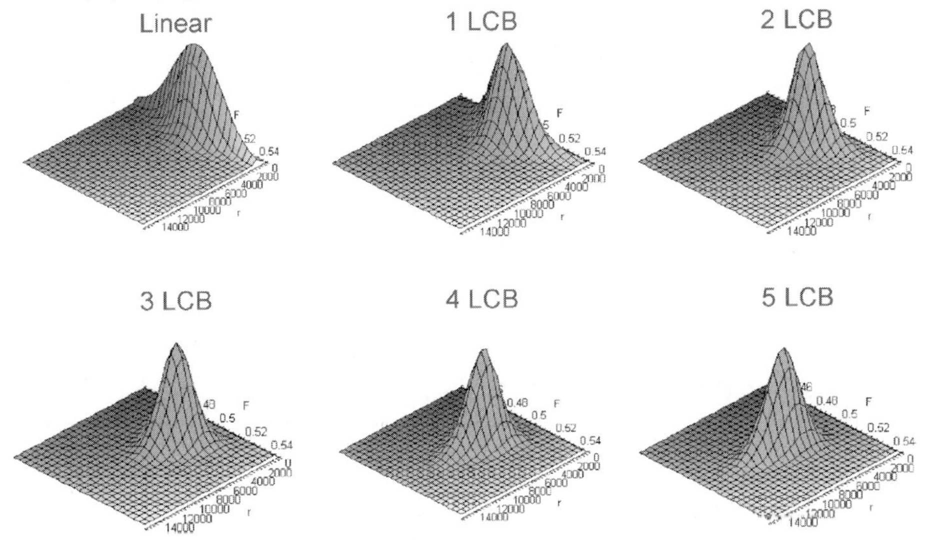

Figure 13.
Trivariate distribution of chain length, chemical composition and long chain branching. Model parameters: $\tau = 0.002$, $\bar{F} = 0.5$, $r_1 r_2 = 1$.

tion for chains with different number of LCBs per chain, as defined by Equation (1). As the number of LCBs per chain increases, the CLD move towards higher averages and the CCD becomes narrower. This equation quantifies the intuitive notion that more branched chain will have higher average lengths and that longer chains will have compositions that are closer to the average copolymer composition. This is, of course, the same prediction given by Stockmayer distribution for linear chains.

Each of the distributions shown in Figure 13 are normalized but, in reality, their fractions in the whole polymer will vary depending on the polymerization conditions.[1,2,12] Regardless of these conditions, the less branched chains always have higher molar fractions (but not necessarily higher mass fractions) than more branched chains.

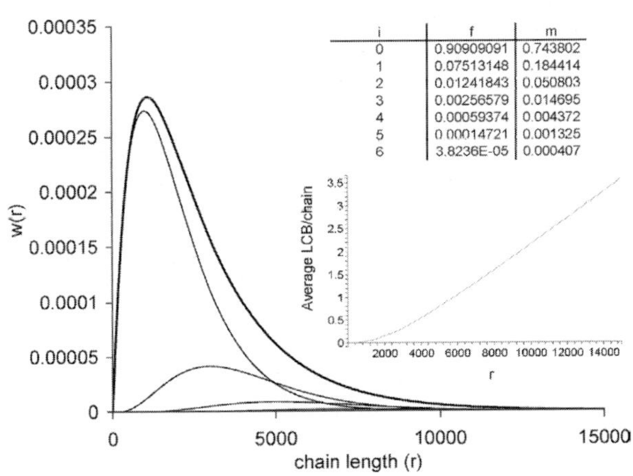

Figure 14.
CLD and average LCB/chain distributions for a model branched polymer. Model parameters: $\alpha = 0.1$ and $\tau = 0.001$.

Molar and mass fractions of populations with i LCBs per chain are given by the equations:

$$f_i = \frac{(2i)!}{i!(i+1)!} \frac{\alpha^i}{(1+\alpha)^{2i+1}} \tag{25}$$

$$m_i = \frac{(2i)!}{i!(i+1)!} \frac{\alpha^i(1-\alpha)(2i+1)}{(1+\alpha)^{2i+2}} \tag{26}$$

The parameter α is a function of polymerization conditions and catalyst type and varies from zero (no LCB formation) to one (infinity LCB formation).[12]

Figure 14 shows the overall and individual CLDs of a model branched polymer. The average number of LCBs as a function of chain length is also illustrated. Notice that most of the polymer chains are linear, but a significant fraction contain LCBs.

It is also interesting to notice how the average number of LCBs per chain increases with chain length. An analytical solution for this distribution is possible, but too lengthy to show here. The final expression is given below:

$$\overline{LCB} = \frac{\phi I_2(2\phi)}{I_1(\phi)} \tag{27}$$

$$\phi = \frac{r\tau\sqrt{\alpha}}{1+\alpha} \tag{28}$$

and I_k are modified Bessel functions of first kind and order k.

Conclusions

In this article we provided an overview of equations that are useful to interpret microstructural distributions of polyolefins. Simple expressions for the distributions of chain length, chemical composition and long chain branching can be found for polymers made with single site catalysts. The treatment of polyolefins made with multiple site catalysts is more elaborate and there is no consensus in the literature on how to best model their microstructure. A simple approach consists in assuming that multiple site catalysts behave as a collection of single site catalysts, but this approach is used for convenience only and is still subject to experimental scrutiny.

We have also shown that these distributions capture the essence of different microstructural characterization analytical approaches, from techniques using single to dual detectors (hyphenated-techniques) and more sophisticated cross-fractionation analyses.

[1] J. B. P. Soares, A. E. Hamielec, *Macromol. Theory Simul.* **1996**, *5*, 547.
[2] J. B. P. Soares, A. E. Hamielec, *Macromol. Theory Simul.* **1997**, *6*, 591.
[3] W. H. Stockmayer, *J. Chem. Phys.* **1945**, *13*, 199.
[4] P. J. Flory, *J. Am. Chem. Soc.* **1936**, *58*, 1877.
[5] Z. Schulz, *Physik. Chem.* **1937**, *B30*, 184.
[6] J. D. Kim, J. B. P. Soares, G. L. Rempel, *J. Polym. Sci.: Part A: Polym. Chem.* **1999**, *37*, 331.
[7] J. B. P. Soares, A. E. Hamielec, *Polymer* **1995**, *36*, 2257.
[8] B. Monrabal, et al., *Macromol. Symp.* this issue.
[9] J. B. P. Soares, S. Anatawaraskul, *J. Polym. Sci.: Part B: Polym. Chem.* **2005**, *43*, 1557.
[10] S. Anantawaraskul, J. B. P. Soares, P. M. Wood-Adams, *Adv. Polym. Sci.* **2005**, *182*, 1.
[11] A. A. da Silva Filho, G. B. de Galland, J. B. P. Soares, *Macromol. Chem. Phys.* **2000**, *201*, 1226.
[12] J. B. P. Soares, *Macromol. Mater. Eng.* **2004**, *289*, 70.

Development of an Automated Cross-Fractionation Apparatus (TREF-GPC) for a Full Characterization of the Bivariate Distribution of Polyolefins

Alberto Ortin, Benjamin Monrabal, Juan Sancho-Tello*

Summary: A compact automated instrument has been developed for measuring the bivariate distribution by TREF fractionation and subsequent GPC analysis of the fractions in a single run. The configuration of this instrument and its operation principles are covered here. High resolution TREF fractionation of HDPE and fast methods with overlapped GPC injections are also discussed. Future developments, such the addition of comonomer or molar mass sensitive detectors, as well as operation in GPC – TREF mode for broad MWD resins are outlined.

Keywords: fractionation of polymers; gel permeation chromatography (GPC); microstructure; polyolefins; temperature rising elution fractionation (TREF)

Introduction

Advances in polyolefin catalyst research and manufacturing processes, such as multiple reactor technologies, have resulted in quite complex end products that can be accurately tailored to different applications by producing carefully designed microstructures in terms of molecular weight distribution, chemical composition distribution, and how the comonomer is incorporated as a function of the molecular weight. This trend will continue in the future and is already challenging the polyolefin characterization technology which needs to account for the different distributions of a given product and for their interdependency. The full characterization polyolefin microstructure can be achieved by defining a two dimensional distribution with molecular weight and chemical composition as independent variables, which constitutes the so-called bivariate distribution.

Single dimension fractionation techniques have been widely used for both determining the molecular weight distribution (MWD) by gel permeation chromatography[1] and the chemical composition distribution (CCD) by temperature rising elution fractionation (TREF)[2,3] or crystallization analysis fractionation (CRYSTAF).[3,4]

In order to elucidate the relationships between structural parameters, different methods have been developed using several online detectors coupled single dimension analytical techniques. Examples of these techniques include the determination of comonomer distribution across the MWD by fixed band[5] or Fourier transform[6] infrared detectors coupled to GPC, or the measurement of molecular weight along the CCD by addition of molar mass sensitive detectors to TREF[7,8] or CRYSTAF[9,10] instruments. Although being powerful tools, those methods cannot provide all the information on polyolefin microstructure, as they are based on fractionation according to one microstructural distribution and the simultaneous measurement of only average values of the other distribution, thus loosing significant information in the process.

The direct measurement of the two-dimensional distribution by cross-fractionation in terms of both molecular weight and

Polymer Characterization, c/ Nicolás Copérnico 10, Valencia Technology Park, 46980 Paterna, Spain
E-mail: info@polymerchar.com

chemical composition is the most comprehensive way to obtain the full definition of polyolefin microstructure. This goal can be achieved by coupling the CCD fractionation by TREF and the MWD fractionation by GPC to measure the bivariate MWD-CCD for the polymer. The method can be further extended by the use of multiple detectors after the cross-fractionation process

Possibly due to the lack of the appropriate instrumentation and high effort required, this comprehensive approach has not been widely used until recently, despite of its potential. Preparative fractionation followed by analysis of the fractions has been the preferred method when trying to elucidate this bivariate distribution, but this procedure is both time consuming and quite complex, taking from days to weeks before the results are available. A description of this method and its application was given by Wild, starting with a preparative fractionation by TREF followed by off-line GPC analysis of the fractions.[11] The first attempt to automate an analytical system for cross-fractionation of polyolefins was presented by Nakano and Goto,[12] who combined a crystallizability fractionation device with a commercial GPC instrument. Recently, an in-house cross-fractionation system has also been described,[13] having increased throughput by using multiple TREF ovens.

A new bench-top, fully automated instrument for cross-fractionation of polyolefins has now been developed based on the combination of a high resolution TREF system (TREF-300, Polymer Char)[9] with a compact dedicated GPC subsystem built as an isothermal oven holding the columns and detectors. The standard procedure to achieve the cross-fractionation of the polymer samples consists of TREF fractionations performed in discontinuous isothermal steps with subsequent on-line GPC analysis of the TREF fractions and IR detection of the polymer. Details of the operation principles, hardware configuration, and data processing are covered in this paper, together with different application examples.

Experimental Part

A schematic diagram of the new cross-fractionation instrument is shown in Figure 1. It was build by modifying the design of a TREF 300 unit (Polymer ChAR, Spain) which incorporates an oven used for sample preparation and a high precision TREF column oven. Other components are: syringe pump, HPLC pump, high temperature isothermal oven (GPC oven) in which the injection valve, a multi-position selection valve and the GPC column set are placed. A dual band IR4 infrared detector (Polymer ChAR, Spain) is used as the concentration detector.

Inside of the TREF oven, a set of five stainless steel vessels with internal filters and magnetic stir bars is used for dissolution of the up to 5 different samples that can be analyzed sequentially. Solvent is added to the vessels through a syringe pump, while the TREF oven is heated typically to 150 °C. Once the polymer sample is fully dissolved, an aliquot is taken from the vessel through its internal filter and loaded into the TREF column by using again the syringe pump, moving the injection valve to the "load" position. Depending on the sample heterogeneity and number of fractions required, typically 1 to 3 milligrams of material are loaded into the TREF column.

The polymer in the TREF column is then crystallized, typically at 0.5 °C/min with no flow, keeping the injection valve in "load" position during the crystallization process. Meanwhile, a stand-by flow of solvent is maintained through the GPC columns. The flow rate is increased, typically to 1.0 mL/min at a pre-defined stabilization time prior to the first injection. Figure 2 shows a typical temperature profile followed by the TREF oven along the full cross-fractionation analysis, where dissolution, crystallization, and stepwise elution are identified.

Once the polymer has been crystallized and the fractions having different crystallinities have been segregated into the TREF column, a discontinuous elution process is followed at increasing temperature steps. At

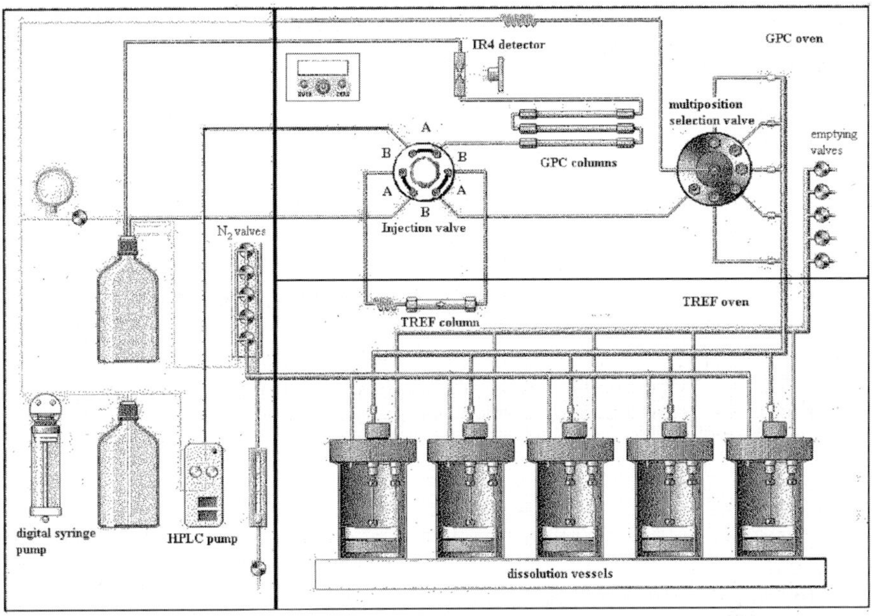

Figure 1.
Schematic diagram of the new automated crossfractionation instrument. Injection valve shown in "load" position A; "inject" position marked as B.

each temperature, and after a given dissolution time, the injection valve is switched to the "inject" position in order to allow the solvent to elute the dissolved polymer from the TREF column. Once that fraction is eluted, the injection valve is closed again to "load" position so that the flow through the TREF column is stopped. Then, the oven temperature is increased to start dissolving the fraction that will be eluted in the next step.

Following that process, different TREF fractions with increasing crystallinity are injected into the GPC columns, where they are fractionated this time according to molar mass. An IR4 infrared detector is used to record the final chromatograms continuously, as depicted in Figure 3, where the raw IR detector and oven temperature signals from a real experiment are plotted. It must be noted that each of the peaks in the IR detector signal is in fact a full GPC

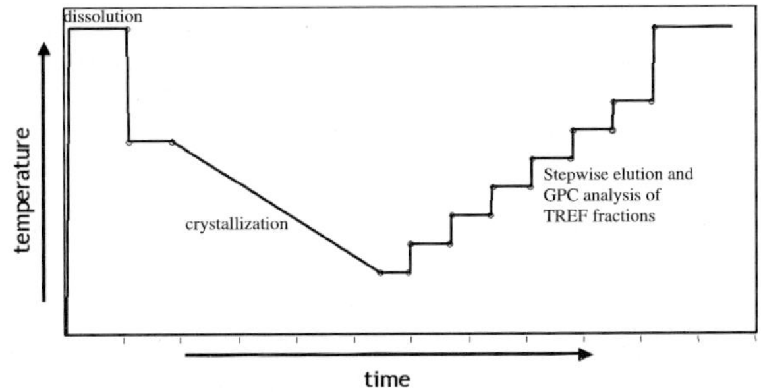

Figure 2.
Temperature profile followed by the TREF oven in a typical cross-fractionation experiment with indication of the different processes.

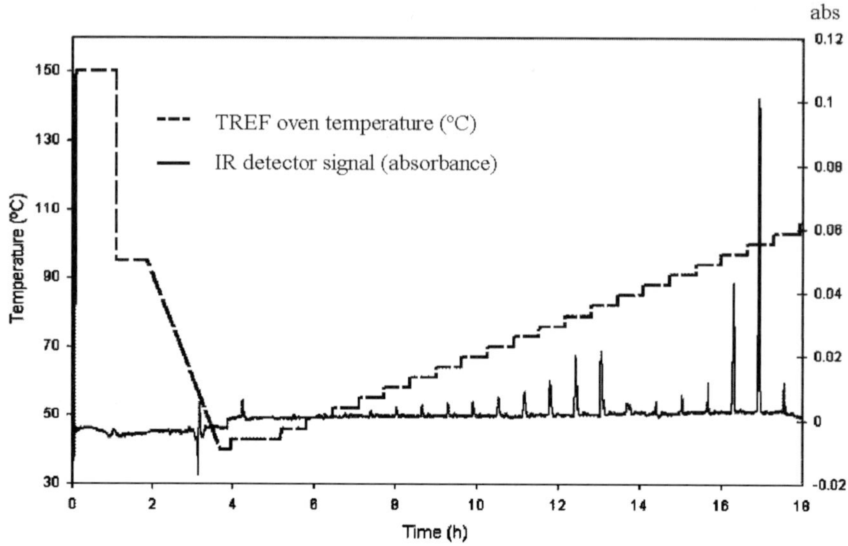

Figure 3.
IR detector signal and TREF oven temperature collected during a full cross-fractionation run.

chromatogram of one of the TREF fractions; therefore the MWD of each narrow-CCD fraction is obtained.

The dissolution and injection processes are illustrated in Figure 4, with an expanded view of the temperature and IR absorbance recorder signals plot. The dashed horizontal arrow indicates the dissolution time of the polymer fraction within the TREF column. The solid downwards arrow shows the injection. Solvent flows through the TREF column for a time long enough to elute all the species dissolved at that temperature. The injection valve is closed as indicated by the solid upwards arrow.

At that time, the TREF oven temperature is increased to that of the next fractionation step. Meanwhile, the TREF polymer fraction travels through the GPC columns, being fractionated by molar mass, producing the peak in the IR detector signal indicated by the circle in Figure 4.

An interesting improvement has been implemented in this system to allow for the injection of one fraction before the previous one has completely exited the GPC columns, hence reducing the overall analysis time. In Figure 5, a direct comparison of standard and overlapped injection analyses is provided. In a period of three hours,

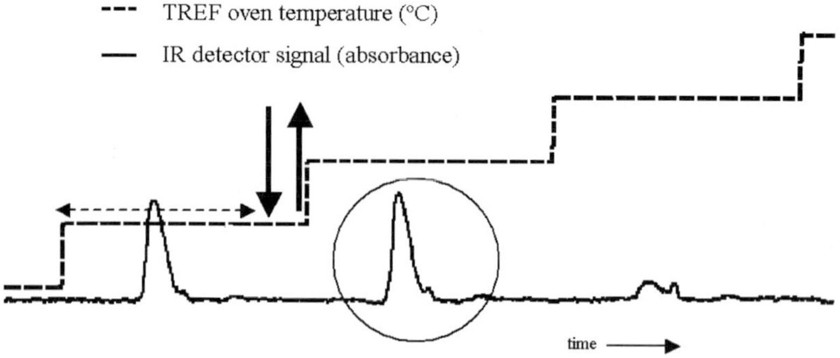

Figure 4.
Detailed view of the injection process.

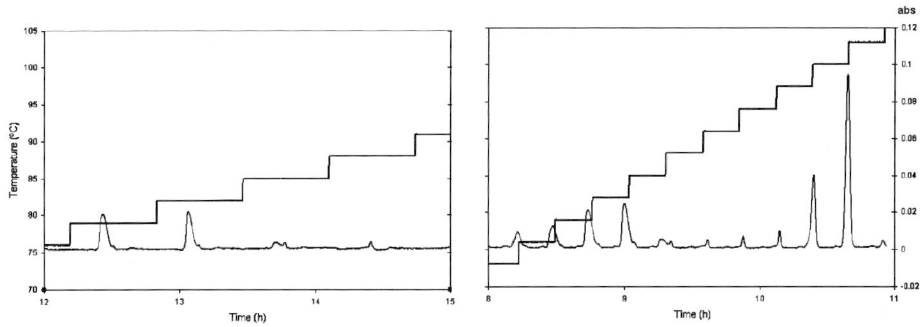

Figure 5.
Comparison of standard GPC injections (left side) versus overlapped GPC injections (right side). In a period of three hours, a larger number of fractions are analyzed in the second case, resulting in shorter analysis time or increased resolution.

only four TREF fractions are analyzed in the first case, while up to eleven fractions can be analyzed when overlapping the injections. This feature can be exploited either to reduce the cross-fractionation time or to increase the resolution of the method by generating more TREF fractions in a given time. No loss of quality in the results was observed by the authors using this technique, probably due to the low concentration of the fractions effectively injected into the GPC columns and also due to the stability of the IR detector along the multiple injection process.

Data Processing

The IR detector generates a continuous signal from which the software automatically separates the individual chromatograms synchronized by the injection signal. This process can be visualized in Figure 6. A series of raw chromatograms, together with their elution temperatures is stored in a raw data file, which is used for further processing.

Every chromatogram is baseline-corrected and integrated in order to obtain the molecular weigh distribution (MWD) based on a polystyrene-standard calibration curve of the GPC columns. Such calibration is generated beforehand by injecting a set of mixtures of narrow distribution polystyrenes to the GPC columns, following the same injection process as for the cross-fractionation analysis, and using the TREF column as injection loop. The different molecular weight averages can be computed after the MWD is acquired, such as the weight, number and z-average molecular weights (M_w, M_n, M_z), as well as

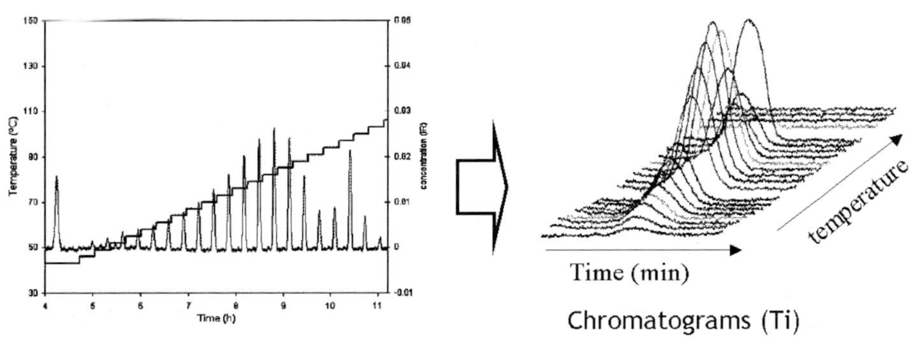

Figure 6.
Continuous raw IR signal (left) and separated chromatograms (right) obtainned as first step in the processing of the cross-fractionation raw signals.

Table 1.
Standard conditions for cross-fractionation analysis of polyethylene samples.

Solvent	1,2 ortho dichlorobenzene with 300 ppm antioxidant (BHT)
Dissolution	60 minutes at 150 °C at 200 rpm stirring inside the stainless steel vessels
Sample weight loaded into TREF column	1.6 mg
Crystallization rate	0.5 °C/min
Elution	24 fractions from 40 °C to 120 °C, temperature step every 3 °C, overlapped GPC injections with 19 minutes inter-injection time
GPC analysis flow rate	1 mL/min
GPC column set	3 PLGel 10 micron mixed columns (Polymer Laboratories Inc.)

polydispersity index (*PDI*). The weight fraction at each temperature is also computed by comparing the area of every single chromatogram to the total area.

That set of MWDs measured at different temperatures are combined with their respective weight fractions to generate a 3D plot representing the bivariate composition-molar mass distribution of the sample. Numerical interpolation smoothing spline algorithms are used in this step, in an approach similar to that described by Nakano and Goto.[12]

The whole sample MWD is also recovered as the weighted sum of the MWD of individual fractions according to their weight fractions. The reconstruction of the TREF profile of the whole sample is a more complex process. It is based on constructing a discontinuous TREF cumulative curve with the progressive sum of the weights of the fractions at each elution temperature, starting with the soluble fraction at the initial temperature and ending with the total weight at the final temperature. The first derivative of the smoothed interpolation line of the cumulative curve constitutes the reconstructed TREF profile.

Results and Discussion

In order to demonstrate the operation of this new cross-fractionation instrument and to show its capabilities, a series of polyethylene samples have been analyzed and the results are presented in this section.

The standard cross-fractionation conditions used are summarized in Table 1, resulting in an overall analysis time of roughly 11 hours per sample for a quite reasonable number of fractions and very good resolution.

Blend Analysis

A two component blend of metallocene polyethylenes was first used to evaluate the resolution of this method. The description of the components in the blend is summarized in Table 2.

A view of the 3D plot of the bivariate distribution of this blend is shown in Figure 7, together with its contour plot. The logM axis was obtained with the GPC calibration curve, while the temperature axis corresponds to the elution temperatures of the different TREF fractions. The elution temperature can be related to comonomer content considering that high crystalline materials eluting at high temperature have little or no comonomer, while increasing levels or comonomer reduce crystallinity and result in lower TREF temperatures. The height of the surface plot represent the relative IR absorbance signal, directly related to concentration at every molar mass-temperature point, so it represents the amount of material having a given molar mass and eluting at a given temperature.

Table 2.
Description of metallocene polyethylene blend components.

Component	Density (g/cm^3)	MI	Weight percent
A	0.957	19.6	50%
B	0.921	0.25	50%

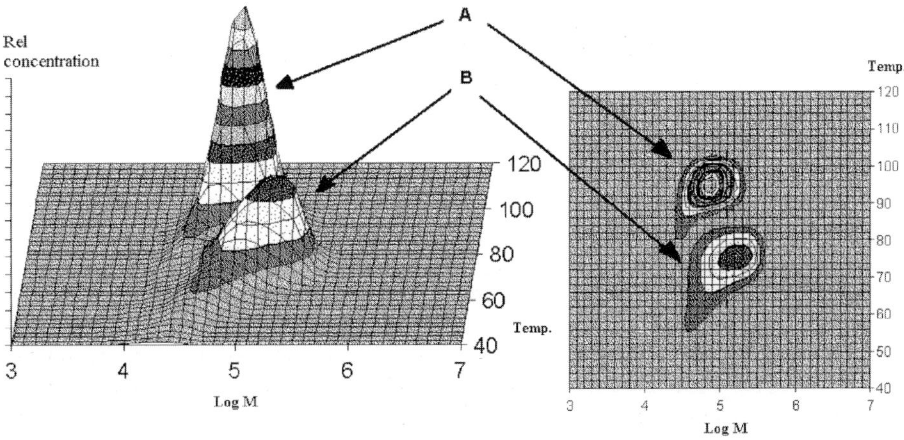

Figure 7.
3D-surface plot (left) and 2-dimensional contour plot (right) of a metallocene polyethylene blend.

The 3D graph on the left side of Figure 7 already gives a clear idea of the separation and of the main structural features of the blend components. Having higher density, component A has an elution temperature around 95 °C, higher than the lower density component B, while its molar mass is shifted to lower values, as expected by their respectives MIs. Component A has narrower composition distribution, resulting in a higher 3 dimensional peak. A certain tailing can also be seen in both components in the direction of lower TREF temperatures and molar masses. This last effect, as well as the relative positions in the temperature-LogM plane are more clearly seen in the contour plot.

The MWD of the fractions, weighted according to their relative amounts, are presented in Figure 8 (thin lines), while the summation of all of them, which constitutes the distribution for the whole sample, is presented with a thicker line in Figure 8.

For comparison purposes, and in order to validate the procedure followed to

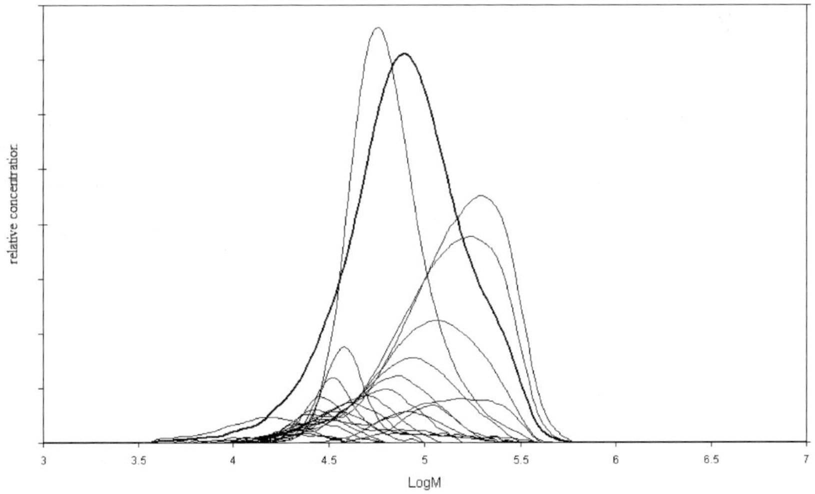

Figure 8.
Weighted MWD curves from all the fractions measured at every elution temperature (thin lines) overlaid with the recovered MWD of the polyethylene blend (thicker line).

recover the original MWD from the fractions, a special analysis was performed in which the whole sample was injected from the TREF column to the GPC columns in the same manner as the fractions were injected. In this way, a direct GPC analysis of the whole sample was performed and the resulting MWD was overlaid in Figure 9 with the recovered MWD. Given the number of individual fractions that were measured and integrated independently, it is remarkable how similar the two results are.

The chemical composition view is obtained by reconstructing the TREF profile from the individually recovered weights of the fractions. This curve is presented in Figure 10, in which the higher resolution achieved in the comonomer dimension is clearly seen, in comparison with the typical lower resolution in the molar mass dimension. The high quality of the recovered TREF curve is quite comparable to that directly obtained with analytical TREF.

Figure 10 also plots the M_w of the fractions as a function of the chemical composition distribution. In this case, the high density material has a lower average molar mass than the lower density one. At the same time, it is also seen that there is a trend of decreasing M_w with increasing comonomer content or lower TREF elution temperatures for both materials.

Similar information can be obtained by coupling molar mass sensitive detectors such as laser light scattering or viscometer detectors to an analytical TREF instrument.[8–10] However, this information is much more evident in the 3D surface and contour plots, which can only be directly measured by this cross-fractionation technique. It is important to emphasize the tremendous power of having the full MWD at each elution temperature, which allows the determination of the polydispersity and other parameters of interest.

Linear Low Density Polyethylene (LLDPE)

A commercial LLDPE was run in this cross-fractionation unit using the same conditions summarized in Table 1. The 3D surface and contour plots are presented

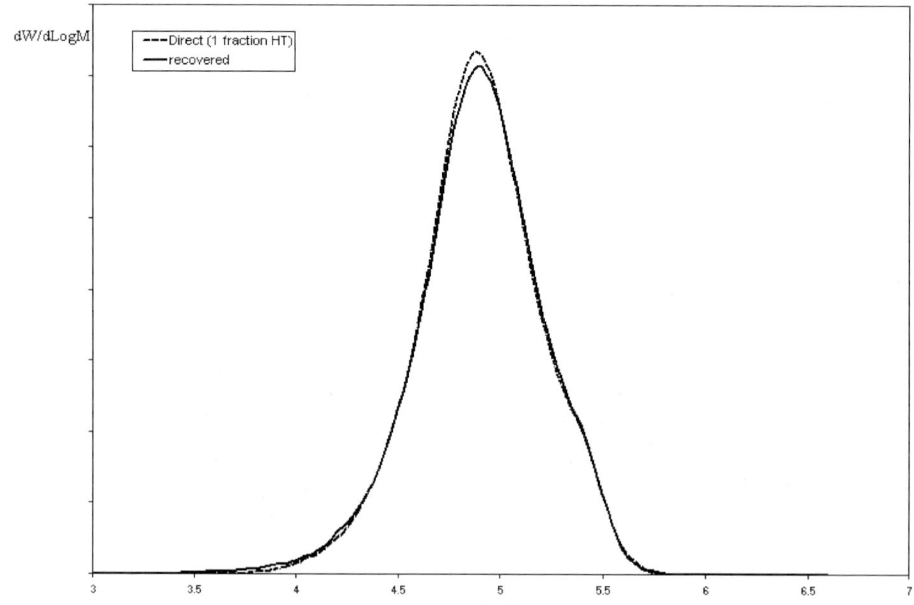

Figure 9.
Overlay of directly measured MWD of the whole polyethylene blend with the recovered MWD after the cross-fractionation process.

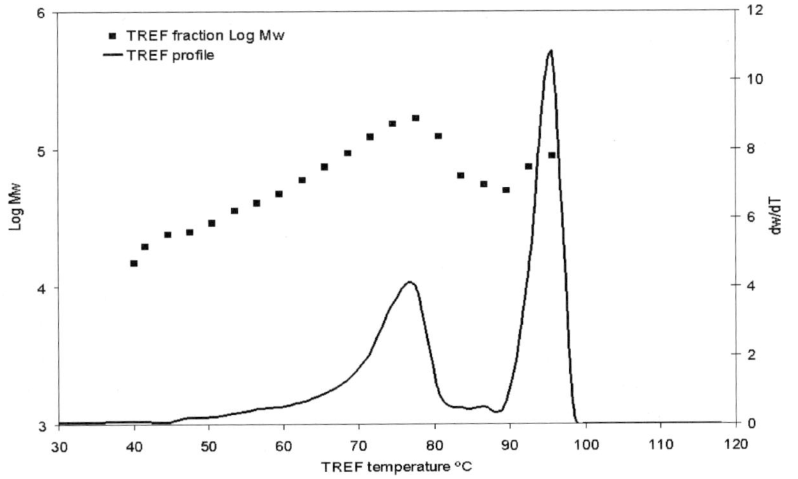

Figure 10.
Reconstructed TREF profile (solid line) based on the weight fractions collected at each elution temperature and weight average molecular weight (M_w) of the individual fractions (dots) as a function of elution temperature.

in Figure 11. In this case, the soluble fraction, amounting to approximately 5 wt% of the sample, is represented as a separate smoothed peak. The contour plot clearly shows the direct dependency of TREF elution temperature and molar mass, in good agreement with previous publications,[11] demonstrating the relative higher comonomer incorporation in the shorter polymer chain, while the less short chain branched fractions (sometimes referred to as homopolymer) exhibits the highest molar mass.

Those observations can be confirmed in the reconstructed TREF view, overlaid with the M_w of the fractions, as presented in Figure 12. We have also overlaid two independent analysis in Figure 12 to show the good reproducibility of this automated analytical instrument and the reliability of the numerical reconstruction of the TREF profiles. The M_w values show more scatter in the temperature ranges with lowest weight fractions because of the lower detector signal to noise ratio in these areas.

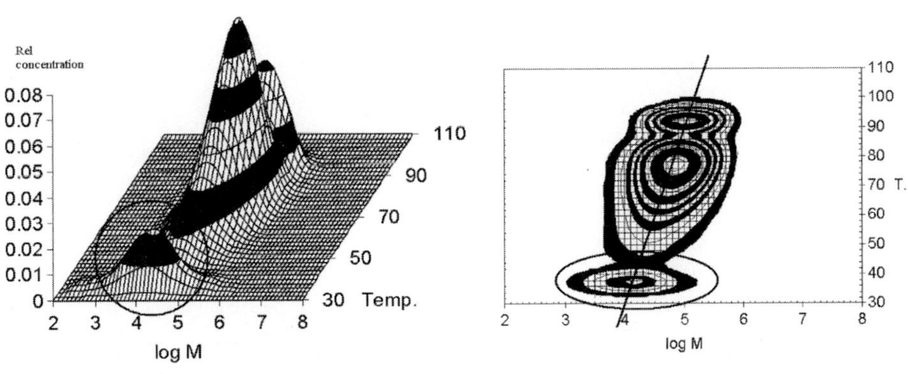

Figure 11.
3D-surface plot (left) and 2-dimensional contour plot (right) of a LLDPE sample including the representation of the soluble fraction as a separate peak. A trend line is added to the contour plot to indicate the direct relation of elution temperature to molar mass.

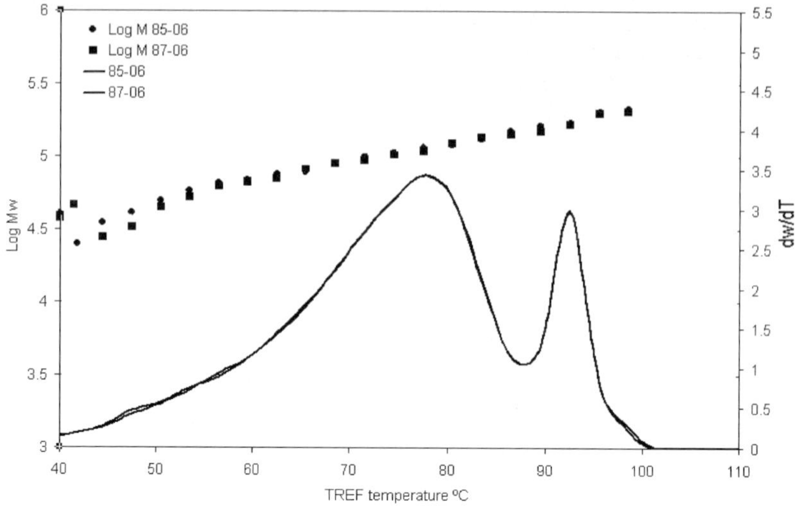

Figure 12.
TREF profile and M_w plot for a LLDPE sample. Two independent cross-fractionation analyses are overlaid to demonstrate the good reproducibility of this technique.

Polyethylene Resins Made in Multiple Reactors

Another commercial product that has a broad and complex chemical composition distribution, in this case produced presumably by a multiple reactor process, was analyzed using the cross-fractionation instrument under the same conditions as in the previous sections. The 3D plots in Figure 13 show a peculiar structure in which three different populations can be identified.

One of the remarkable characteristics of this product is the highly short chain branched component eluting in the range of 55 to 70 °C showing a higher molecular weight than that of the more crystalline population eluting at higher temperatures. This product is a good example of materials with very heterogeneous and multimodal chemical composition distributions. This can also be seen very clearly the reconstructed TREF profile shown in Figure 14.

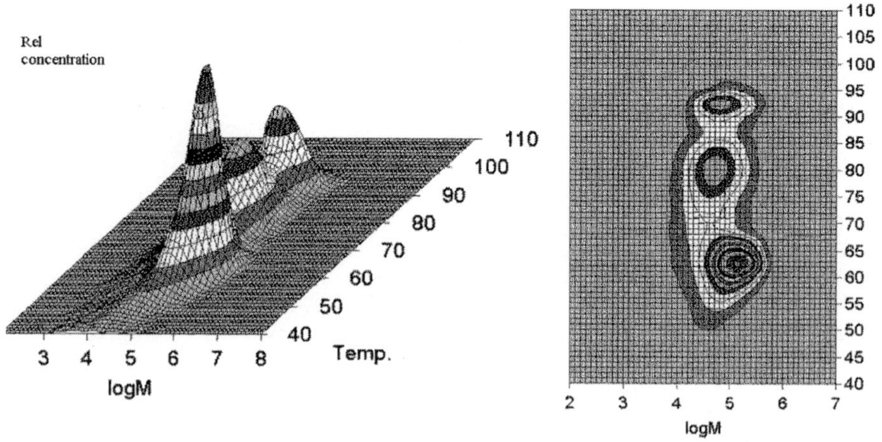

Figure 13.
3D-surface plot (left) and 2-dimensional contour plot (right) of a multiple reactor polyethylene sample.

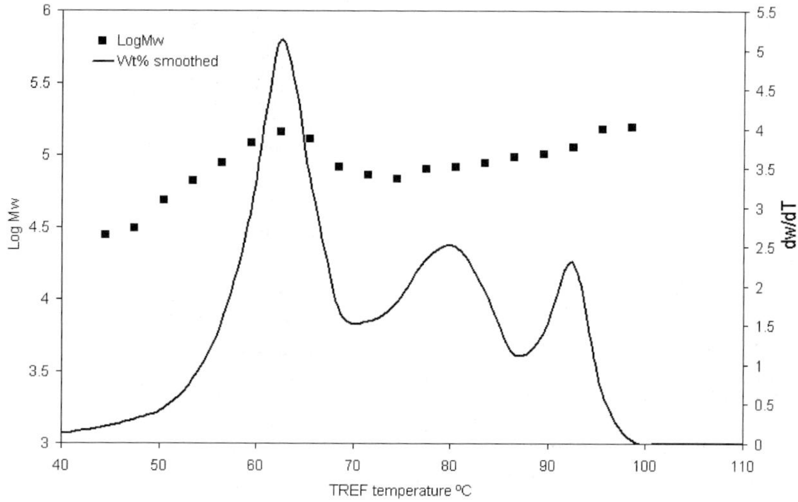

Figure 14.
TREF profile and M_w plot for a multiple reactor polyethylene sample.

On the other hand, the MWD analysis of the whole sample measured by GPC would provide little information on its structural complexity. In fact, the MWD of this product is basically unimodal, showing a small high molecular weight shoulder, as can be visualized from the surface and contour plots of Figure 13 due to the overlapping of the molecular weight distributions of the different polymer populations.

Distribution of Comonomer as a Function of MWD of HDPE Pipe Grade Resins

A higher resolution method was used for the analysis of high density resins because of their narrow chemical composition distribution due to the low amount of comonomer present in the copolymer. At the same time, a very good definition of the bivariate distribution in the 3D plot was required to obtain all the comonomer distribution information. Analysis conditions are summarized in Table 3.

In the lower temperature range from the 60 °C to 82 °C, TREF fractions were taken every 2 °C so that enough material was eluted to produce a detectable chromatogram. Fractions were taken every 1 °C from 82 °C to 100 °C in order to obtain the best temperature resolution and gather the maximum information from the cross-fractionation process.

Figure 15 presents the 3D surface and contour plots after the high resolution cross-fractionation analysis of a pipe grade

Table 3.
High resolution conditions for cross-fractionation analysis of HDPE samples.

Solvent	1,2 ortho dichlorobenzene with 300 ppm antioxidant (BHT)
Dissolution	60 minutes at 150 °C at 200 rpm stirring rate inside stainless steel vessels
Sample weight loaded into TREF column	1.6 mg
Crystallization rate	0.1 °C/min
Elution	27 fractions from 60 °C to 105 °C, temperature step every 1-2 °C, overlapped GPC injections with 19 minutes inter-injection time
GPC analysis flow rate	1 mL/min
GPC column set	3 PLGel 10 micron mixed columns (Polymer Laboratories Inc.)

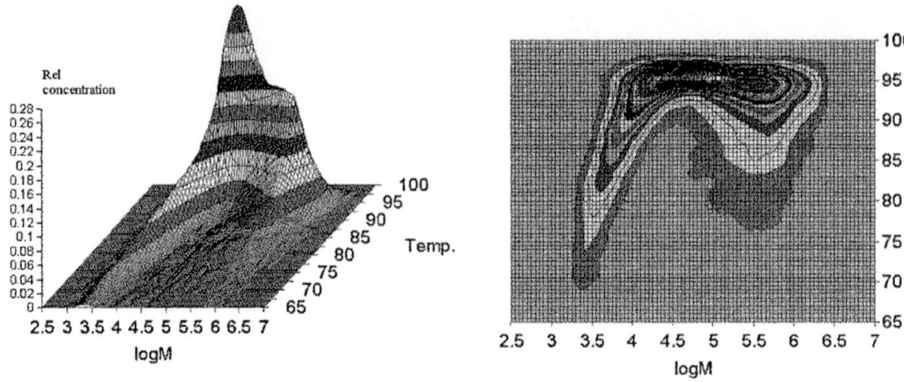

Figure 15.
3D-surface plot (left) and 2-dimensional contour plot (right) of a HDPE pipe grade resin.

HDPE sample. This HDPE is a reverse comonomer composition resin, in which a low level of comonomer was added to the higher molecular weight components in order to achieve the desired end product properties.

Figure 15 clearly shows the heterogeneity of this product. The pronounced tailing towards lower elution temperatures in the low molecular weight range is explained by the effect of the chain ends which act as a source of crystal defects as described by Nieto et al.[14] and produce a depression in TREF elution temperatures. The fact that for molecular weights lower than 10,000 there is a rather sharp decrease in TREF temperatures is also in good agreement with Nieto's observations based on CRYSTAF analysis of narrow-MWD polyethylene fractions and paraffins. This low molecular weight effect on TREF elution temperatures can be compensated in the processing software when translating the TREF temperature axis into the comonomer weight percent in order to produce the two dimensional distributions of molar mass and comonomer composition.

On the other hand, the higher molar mass fractions from Log M from 5 to 6 show a broad TREF elution temperature range produced, in this case, by the addition of comonomer in that area of the molar mass distribution.

In the TREF view plot (Figure 16), M_w values are rather uniform for most all temperatures, decreasing only at lower temperatures because of the low molar mass tail observed in the two dimensional plots. The different contributions to the TREF temperature decrease by either comonomer addition or low molar mass species are clearly seen in this two-dimensional plot measured with the cross-fractionation instrument. They would not be detected by one-dimensional techniques, even with multiple detectors.

Figure 16 also shows how the polydispersity index depends on the TREF elution temperature. PDI increases dramatically as the temperature is reduced, from initial values of approximately 5 at high temperatures to about 30 at low temperatures, due to the bimodality of the measured chromatograms.

In the case of these HDPE products with broad MWD it has been of great interest to study the molar mass view and also to add the comonomer distribution information to it. An approach to investigate this problem using the cross-fractionation information in the 3D plots is outlined in the following paragraphs.

The recovered MWD for the whole sample is shown in Figure 17. The MWD is clearly bimodal and the question now is to determine the range across this broad MWD in which comonomer is present.

In this study we start by realizing that for every molar mass value, a TREF-like profile is obtained from the 3D surface plot,

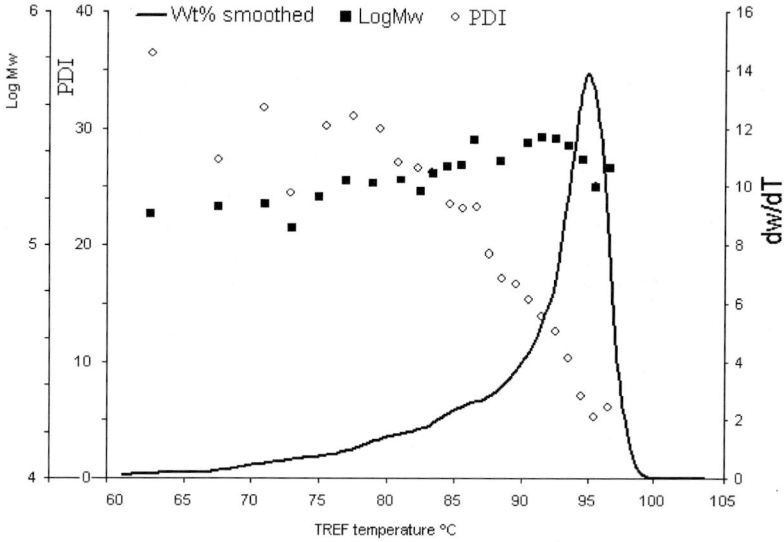

Figure 16.
TREF profile and M_w and PDI plots for a HDPE pipe grade resin.

representing the TREF profile of a very narrow molecular weight fraction. Figure 18 helps visualizing this approach by presenting a surface plot cut in the direction of the TREF temperature axis, highlighting the equivalent TREF curve at log M value of 5.5.

For the generated TREF curve, the weight average elution temperature, T_w, is computed with the equation:[4]

$$T_W = \frac{\sum c_i \cdot T_i}{\sum c_i} \quad (1)$$

Similar TREF curves and T_w averages can be obtained for every molar mass value to generate a T_w versus Log M plot. The

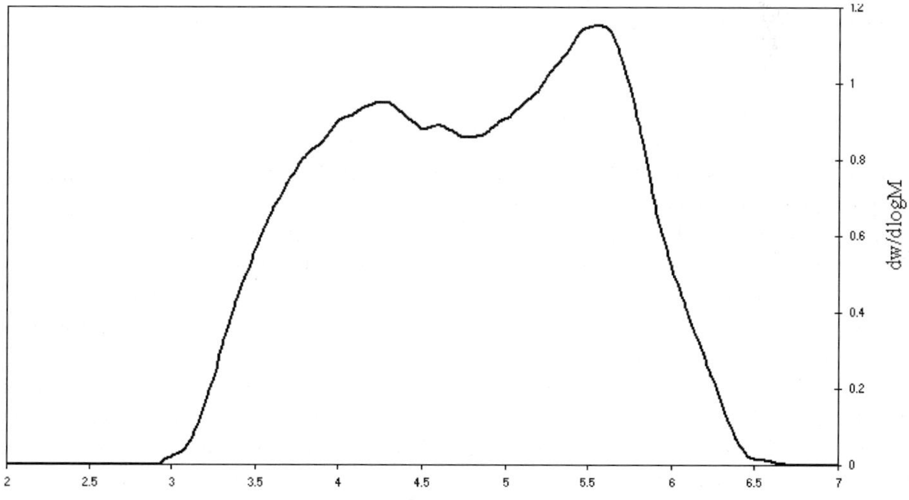

Figure 17.
Recovered MWD of the whole HDPE sample by a weighted sum of the individual MWDs based on their respective weight fractions.

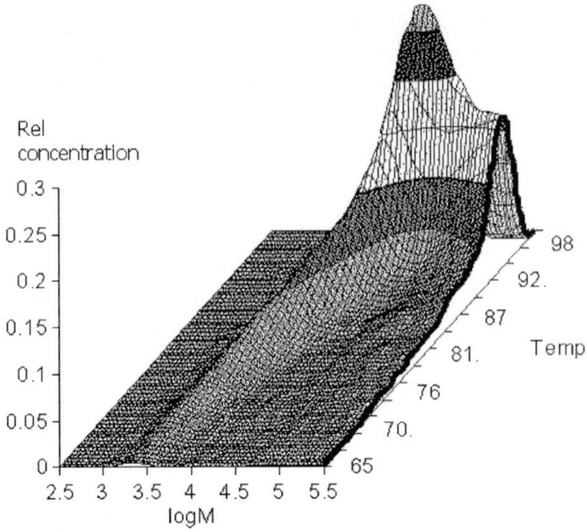

Figure 18.
3D surface plot of a HDPE sample with a cut view produced at LogM = 5.5 in the direction of the temperature axis. The highlighted edge of the surface plot represents a TREF profile at that molar mass value.

resulting data series is overlaid with the MWD curve in Figure 19.

It is well established[2,3,15] that T_w is correlated to the methyls groups per 1000 carbons (CH_3/1000C), even for molecular weights down to 1000 (either coming from short chain branches or chain ends) by a linear inverse function; consequently, a

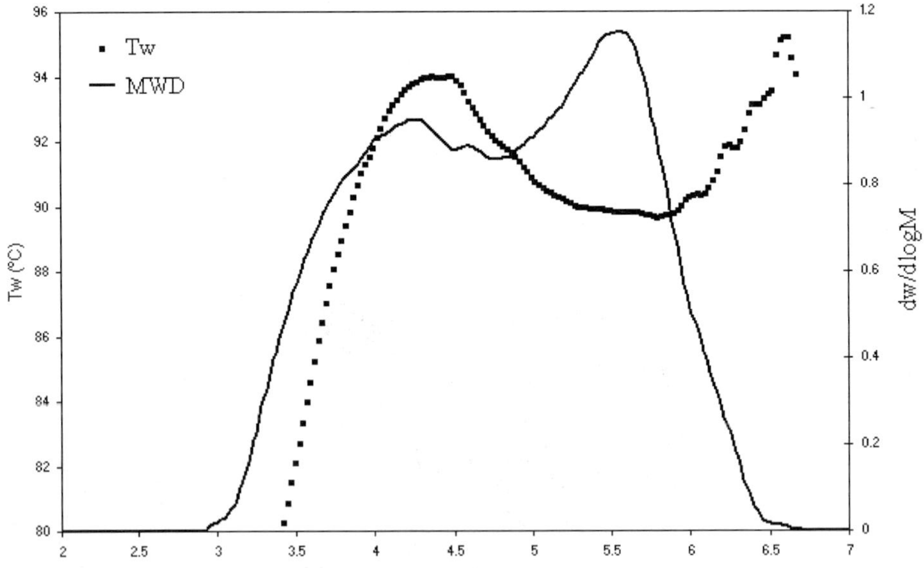

Figure 19.
MWD of a HDPE sample and T_w averages as a function of molecular weight. The values of T_w decrease sharply for molecular weights lower than 10,000 due to chain end effects in TREF fractionation. T_w decreases in the high molecular weight range due to comonomer incorporation.

linear calibration curve can be developed to translate T_w values into CH_3/1000C. Although work is being conducted to obtain an accurate calibration curve for this method, in order to show the potential of this approach, a simulated T_w –CH_3/1000C calibration curve was used in this work to obtain the methyl frequency as a function of molecular weight.

A correction for the effect of chain ends based on molecular weight was applied to obtain the short chain branch frequency (SCB/1000C) along the MWD (Figure 20). Provided the type of comonomer is known, as it is usually the case, the comonomer weight percent can be directly computed from the SCB frequency.

This result proves that the cross-fractionation instrument operating as TREF–GPC can provide the same information as online or offline GPC-IR techniques. A higher precision is expected from the cross-fractionation method because precision is not related to detector signal-to-noise ratio but to the TREF separation mechanism which is known to be highly reproducible.

An alternative approach, starting with fractionation by molecular weight and then analyzing the narrow-MWD fractions by TREF may seem more suitable for those broad MWD samples. However, provided that the bivariate distribution is accurately measured, there should be no difference in using TREF or GPC as the primary fractionation method, and the choice of cross-fractionation configuration should be made according to convenience and available instrumentation.

Conclusions

A new automated compact cross-fractionation instrument has been developed with special attention to efficiency, flexibility and robustness. The main hardware components, as well as operation principles, have been described in some detail, together with a special feature for producing overlapped GPC analysis of TREF fractions.

A series of application examples has been presented for different types of

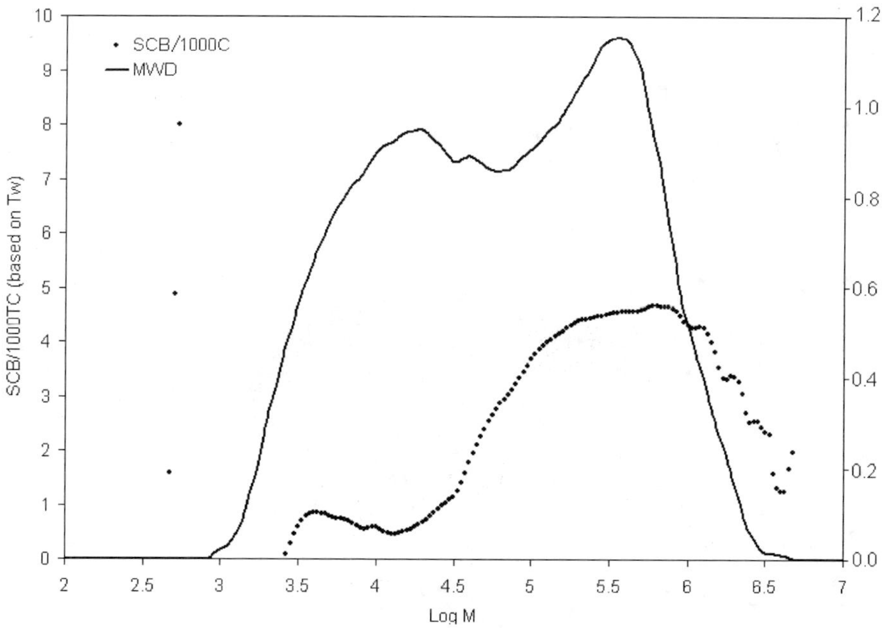

Figure 20.
Short chain branch (SCB) frequency along the MWD obtained from the cross-fractionation data, showing that the higher molecular weight material has the higher SCB frequency due comonomer incorporation.

polyethylene samples to demonstrate the resolution and comprehensive information produced by this instrument. Other analytical methods with slower crystallization rates down to 0.01 °C/min, sub-ambient temperatures for less crystalline polymers, higher elution temperatures for polypropylene, or different flow rates are easily programmed through the virtual instrumentation software user interface.

This cross-fractionation instrument also proves to be a very powerful tool for obtaining the SCB distribution along the MWD for the most demanding applications of pipe grade HDPE resins having very little comonomer incorporated. An approach for measuring the distribution of comonomer along the MWD for HDPE products based on cross-fractionation data obtained by TREF fractionation and GPC analysis has been studied and its feasibility established using the instrument and analysis conditions described in this work.

Addition of other detectors online will still increase the amount of information generated by this cross-fractionation instrument. A methyl sensitive head can be incorporated to the IR detector in order to measure the methyl groups concentration and, as a result, the comonomer weight percent directly both for polyethylene and polypropylene copolymers. Future developments also include coupling viscometer and light scattering detectors to this instrument, which will provide absolute molecular weight and long chain branching detection as well as the possible adaptation to work in reverse mode as GPC fractionation followed by the TREF analysis of the GPC fractions.

[1] W. W. Yau, J. J. Kirkland, D. D. Bly, "*Modern size exclusion liquid chromatography*", Wiley, **1979**.
[2] L. Wild, T. Ryle, D. Knobeloch, I. R. Peat, *J. Polym. Sci., Polym. Phys. Ed.* **1982**, *20*, 441.
[3] B. Monrabal, Polymer ChAR, in: "*Encyclopedia of Analytical Chemistry*", R. A. Meyers, Ed., John Wiley &Sons Ltd., **2000**.
[4] B. Monrabal, in: "*New Trends in Polyolefin Science and Technology*", S. Hosoda, Ed., Research signpost, **1996**, 126.
[5] J. V. Bosch, A. Ortin, B. Monrabal, International GPC Symposium proceedings, Arizona, **1998**, 633–640
[6] P. J. DesLauriers, D. C. Rohlfing, E. T. Hsieh, *Polymer* **2002**, *43*, 159.
[7] L. G. Hazlitt, D. G. Moldovan, US Pat. 4,798,081 (**1989**).
[8] W. W. Yau, D. Gillespie, *Polymer* **2001**, 42, 8947.
[9] B. Monrabal, A. Ortin, L. Romero, 12th *Intern. Symp. on Polymer Analysis and Characterization,* La Rochelle June 28 **1999**.
[10] B. Monrabal, A. Ortin, P. del Hierro, L. Romero, *International GPC Symposium proceedings*, Las Vegas, **2000**.
[11] L. Wild, T. Ryle, D. Knobeloch, *Polym. Prepr., Am. Chem. Soc., Polym. Chem. Div.* **1982**, *23*, 133.
[12] S. Nakano, Y. Goto, *J. Appl. Polym. Sci.* **1981**, *26*, 4217.
[13] C. Li Pi Shan, D. Gillespie, L. Hazlitt, Ecorep **2005** Lyon.
[14] J. Nieto, T. Oswald, F. Blanco, J. B. P. Soares, B. Monrabal, *J Polym Sci. B: Polym. Phys.* **2001**, *39*, 1616.
[15] B. Monrabal, F. Blanco, J. Nieto, J. B. P. Soares, *J Polym Sci. B: Polym. Chem.* **1999**, *37*, 89.

 DOI: 10.1002/masy.200751103

Examples of Using 3D-GPC-TREF for Polyolefin Characterization

*Wallace W. Yau**[1,2]

Summary: Selected examples of using 3D-GPC-TREF to solve polyolefin characterization problems are described in this paper. The term 3D-GPC-TREF stands for a home-build hybrid system of gel permeation chromatograph (GPC) coupled with the capability of the temperature rising elution fractionation (TREF) that includes three online detectors, i.e. the infrared (IR), the differential-pressure viscometer (DP), and the light scattering (LS) detectors.

Keywords: polyolefin characterization; short-chain and long-chain branching; structure-property relations; triple-detector GPC; triple-detector TREF

Introduction

A hybrid 3D-GPC-TREF system is built by installing a TREF add-on oil bath to an existing Waters 2000 CV GPC system[1] that has a built-in refractive index detector and a differential-pressure viscometer. Two additional detectors are added to the system; these are the Polymer ChAR IR4 dual-wavelength detector[2] and the PDI-1040 light scattering detector.[3] The Polymer Lab 20 micron mixed bed light scattering columns[4] are used in the GPC experiment. The configuration of the instrument is shown in the schematic in Figure 1. The six port valve in the system automatically switches the hot solvent flow through either the GPC or the TREF columns at several pre-determined time sequence steps. The column selection valve-switching is automatically synchronized with the oil bath temperature cooling and heating cycles in the TREF mode of operation. With the auto-sampler capability of the Waters 2000 CV system, up to 24 TREF samples can be loaded at one time and analyzed in an unattended operation at the rate of eight samples for every 24 hours. The TREF conditions for most of results reported in this paper are the following: the solvent is o-dichlorobenzene (ODCB) at 150 °C; the TREF column is packed with 27 micron glass beads; each three-hour TREF cycle consists of a 1.5 °C/min cooling step down to 25 °C followed by a 2 °C/min heating step back to 150 °C. The operation of this hybrid system has been reported in detail in two earlier publications.[5]

Polyolefin samples can have complex microstructures due to polymer branching and stereo-regularity differences. While the 3D-GPC experiment can be used to study polymer molecular weight distribution (MWD) and long-chain branching (LCB), the 3D-TREF experiment is needed to analyze polymer short-chain branching (SCB), co-monomer composition distribution (CCD), and the effect of stereo-regularity differences. For example, in a sample of Ziegler-Natta linear-low-density polyethylene (ZN-LLDPE), multiple TREF peaks are commonly observed that reflect the multiple-site nature of this catalyst type. However, the MWD curves of these ZN sites are broad and overlapping; therefore, they are not resolved in the 3D-GPC experiment. This is the value of using

[1] Equistar Chemicals, a Lyondell Company, 11530 Northlake Drive, Cincinnati, Ohio 45249 USA
E-mail: wallace.yau@equistarchem.com
[2] Present address: Characterization Group, Plastics R&D, B1470-D131, The Dow Chemical Company, 2301 Brazosport Blvd., Freeport, Texas 77541
E-mail: wyau@dow.com

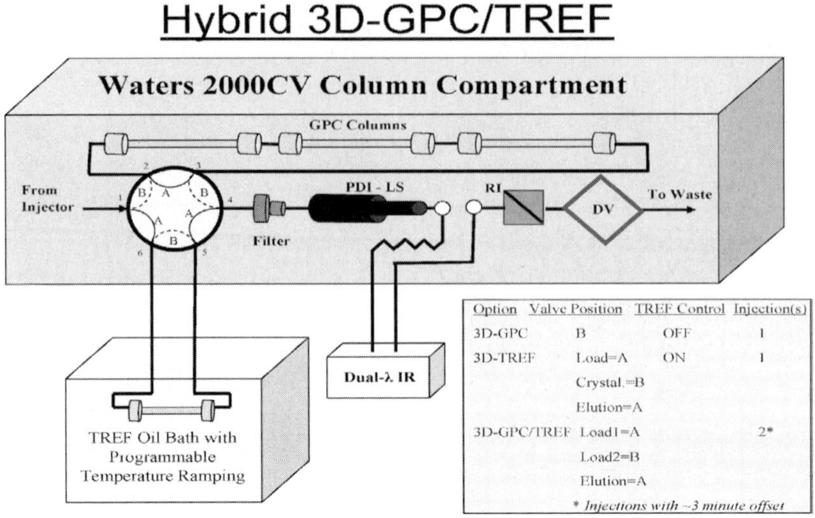

Figure 1.
Configuration of the hybrid 3D-GPC-TREF system.

GPC and TREF in a complementary way to analyze polyolefins for greatly improving the understanding of their complex structures.

One way to explain the complex multiple-site nature of polyolefin structures is to visualize it in a three-dimensional picture, as shown in Figure 2. In this illustrative model of a ZN-LLDPE sample, we can explain the structural features seen by GPC versus those seen by TREF. When viewing the graph in the GPC direction (bottom left frame), one sees the result of the overlapping MWD profiles of the multiple sites that are stacked together with one behind

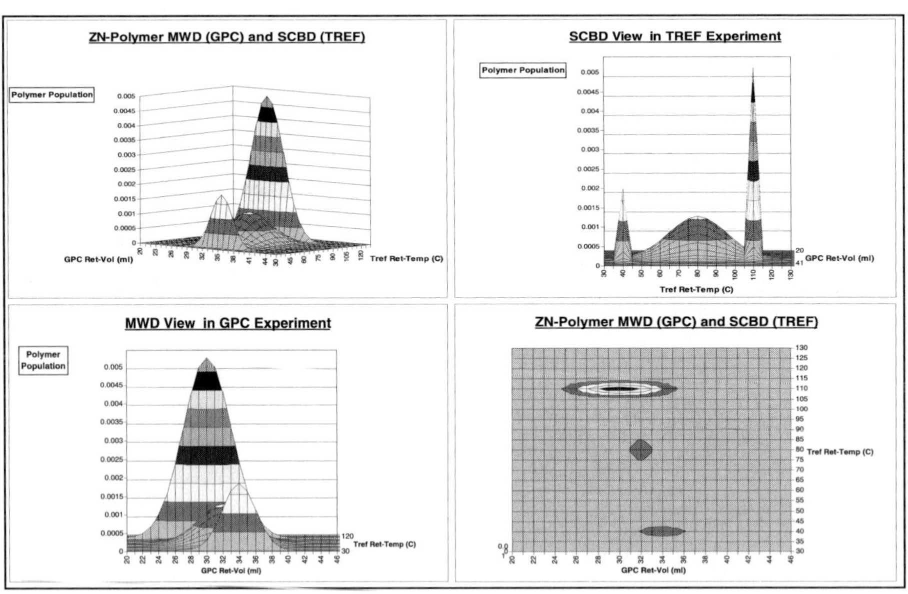

Figure 2.
A three-dimension illustration of the complex nature of polymer μ-structure.

the other. Only in the TREF view (top right frame) can one see the resolution of the multiple-site populations in the polymer structure. Viewing from the top (bottom right frame), one sees the three-dimensional aspect of the polymer structure.

In the study examples presented in the next section, we will see the synergistic value of combining the 3D-GPC and the 3D-TREF results in solving polymer characterization problems. In some cases, we show a single "polymer μ-structure" plot made up by the side-by-side display of the results from 3D-GPC and 3D-TREF. The term "polymer μ-structure" is used in reference to the molecular structural features of polymer that include MW, MWD, SCB, LCB, and CCD.

Study Examples

There are many unique features and capabilities of the 3D-GPC-TREF technique useful for supporting all functions in a polyolefin R&D program. They have been used to solve problems in catalyst research, pilot plant and product development trials, tech-service and customer complaints, and competitive product analyses. A selective number of examples of such studies are presented below.

High Precision Detection of Subtle Polymer Structural Differences

In an effort to develop a LLDPE product to qualify for a stretch film application, some level of LCB detrimental to film performance was detected by 3D-GPC as shown in Figure 3. The level of LCB can be quantified by the size of the high molecular weight hump in the LS curves. The result shows that the LCB is much less in the two competitor products. It is plausible that this unexpected LCB in the current product might be the result of the melt extrusion process used to produce polymer pellets. To answer this question, tests were done on a sample of the reactor powder as well as on the pellets made from the powder. The small LCB difference between the pellet and powder seen in Figure 3 clearly shows that the melt extrusion process could not be the culprit of the LCB problem. This result helped direct attention to the reactor kinetics and catalyst preparation to solve the problem. This example is selected to emphasize the importance of the high precision quality of the polymer analysis data in solving real life problems.

Another example for the need of high precision data is the study of peroxide type in a low-density polyethylene (LDPE) process. As shown in Figure 4, the differences made by peroxide change are very

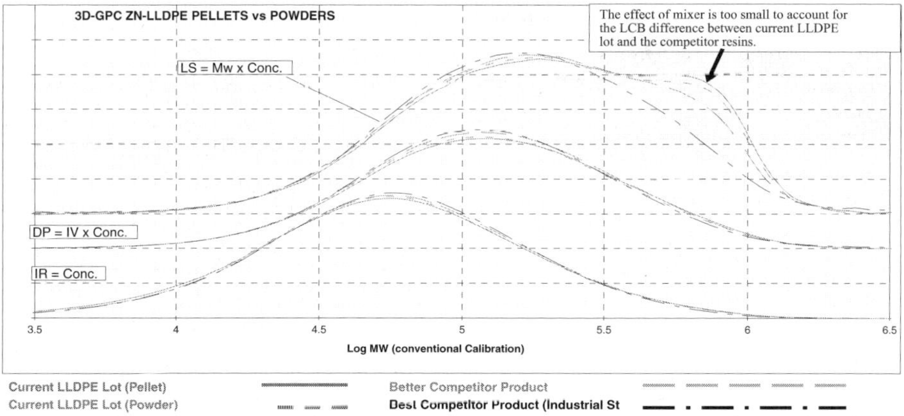

Figure 3.
High precision detection of LCB in ZN-LLDPE.

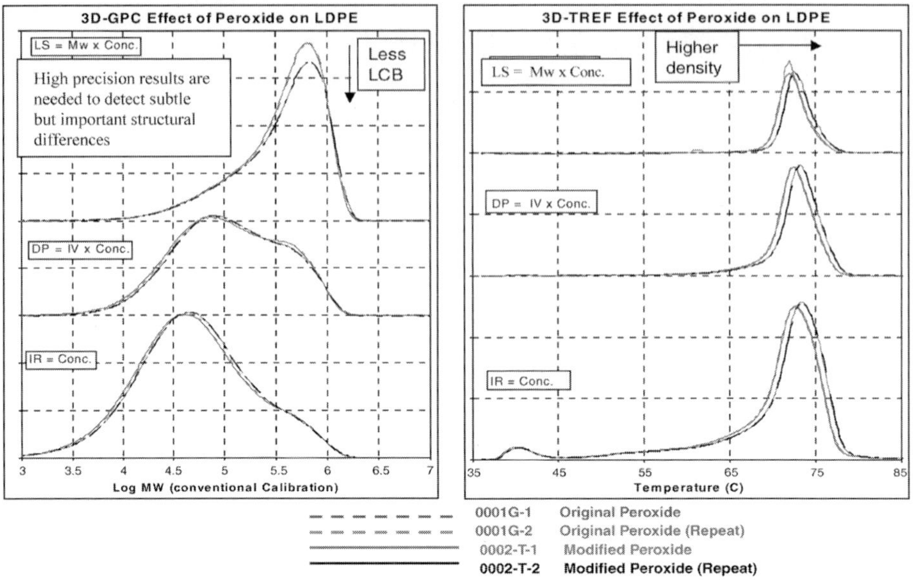

Figure 4.
High-precision 3D-GPC-TREF study of peroxide effect on LDPE production.

small, but they were clearly recognizable in our 3D-GPC-TREF results. These subtle LCB differences would not have been revealed so clearly without the high precision features of our 3D-GPC-TREF instrument. The good repeatability of our instrument is shown by the results of repeat sample analyses in Figure 4. These differences in LCB and product density as detected by 3D-GPC-TREF are very small, but they are responsible for significant differences in polymer end-use properties. Great emphasis was made in our home-made 3D-GPC-TREF system with many built-in automated control features to maximize the repeatability of sample analyses.

LCB Detection in 3D-GPC-TREF

As demonstrated in Figure 3 and 4, the effect of LCB can be seen qualitatively by visual inspection of the 3D-GPC or 3D-TREF curves. However, in order to quantify the LCB level more precisely, we have introduced two LCB indices to process the data, as described in Figure 5 and 6. The formulations of these two new LCB indices (gpcBR and trefBR)[6] are created in a way that they can be used for direct comparison with the LCB index (LCBI) used in the rheological tests.[7] The definition of gpcBR and trefBR takes advantage of the four most precisely measured parameters in the 3D-GPC-TREF experiments, i.e., the LS-measured M_w, the viscometer-measured IV, and the conventional GPC-measured M_w and M_v values, where M_w, IV, and M_v stand for the weight-average molecular weight, intrinsic viscosity, and viscosity-average molecular weight, respectively. The calculated gpcBR and trefBR values represent the average LCB level in the bulk sample.

The calculated gpcBR and trefBR values are included in the final report of every 3D-GPC-TREF analysis. An example of such a report is shown in Figure 7 and 8 for the comparison of a tubular versus an autoclave LDPE sample. It is commonly observed that the tubular LDPE has higher density (higher TREF elution temperature) and lower LCB (lower gpcBR and trefBR values) than its autoclave LDPE counterpart.

The value of the combined information from 3D-GPC and 3D-TREF serves nicely for the purpose of answering a usual question of "gel" or "un-melt," a problem

gpcBR from 3D-GPC

$$gpcLCB = \frac{K \times M_{w,b}^{\alpha}}{\left(M_{w,L}/M_{v,L}\right)^{\alpha} \times [\eta]_b} - 1 = \left(\frac{M_{w,b}}{M_{w,L}}\right)^{\alpha} \times \frac{[\eta]_L}{[\eta]_b} - 1$$

where,

$[\eta]_b \equiv bulk\ [\eta]\ by\ GPC-Viscometry.$

$M_{w,b} \equiv bulk\ M_w\ by\ GPC-LS.$

$M_{w,L} \equiv conventional\ GPC-M_w.$

$M_{v,L} \equiv conventional\ GPC-M_v\ from\ K\ and\ \alpha.$

$K\ and\ \alpha \equiv Mark-Houwink\ parameters.$

gpcBR = [Mw(LS)/MW(conv)]^0.73 * IV(conv)/IV(vis) – 1

Figure 5.
High-precision LCB index gpcBR by 3D-GPC.

often encountered in thin films of high-density polyethylene (HDPE). The results of one of these studies are shown in the μ-structure plot in Figure 9, where the results of the "gel" particle are being compared with the results of the clear HDPE film base that contains no "gel." Here, the 3D-GPC side of the plot shows the structure being mainly of a linear high molecular weight material with no significant LCB. This observation is derived from the fact that the size of viscosity peak also increases along with the increase in the LS peak. *The presence of LCB would increase the viscometer peak to a lesser extent than the LS peak.* The 3D-TREF side of the plot shows the structure being mainly of the high melting type, just like the HDPE film base, with no indication of significant branching structure. These results are indicative of the problem being more of a high molecular weight "un-melt" rather than that of a "gel" or cross-linking branching problem in the usual sense. Because a very limited amount of the "gel" particles can be extracted from a film sample, the very small

Defining trefBR as a LCB – screening parameter in 3D – TREF (and 3D - Batch) experiments:

$$trefBR \equiv \left[\frac{K * Mw_{LS}^{\alpha}}{[\eta]_{GPCV}}\right] - 1$$

where: $\alpha \approx 0.73$, $K \approx 0.00374$ for polyethylene

Note:

- trefBR = gpcBR without a small polydispersity correction

$$gpcBR = (trefBR + 1) * \left[\frac{Mv_{cc}}{Mw_{cc}}\right]^{\alpha} - 1$$

- Level of polydispersity corrections:
  for broad MWD samples ≈ 0.2 (Pd ~ 10 - 20)
  for narrow MWD samples ≈ 0.1 (Pd ~ 2 - 5)

Figure 6.
High-precision LCB index trefBR by 3D-TREF.

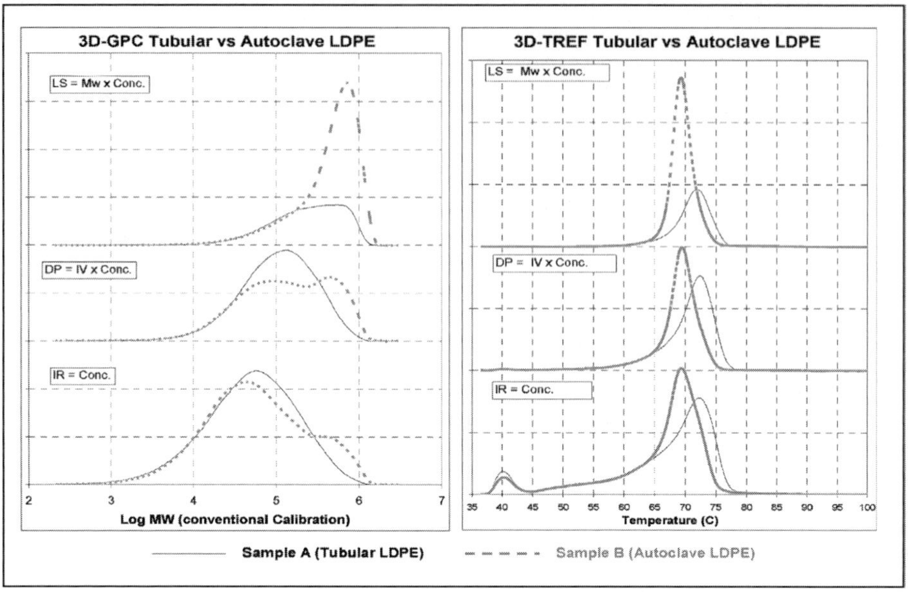

Figure 7.
The 3D-GPC-TREF μ-structure plot for tubular versus autoclave LDPE.

3D-GPC Summary
Job : Tubular vs Autoclave LDPE — *Machine :* **GPC-xxx**

Sample Num	Product #							
Sample-A	Tubular LDPE	**M bulk**	**Mz+1**	**MLS/Mcc**	**MLS/Mncc**	**IV bulk**	**g-prime**	gpcBR
		190,404	574,868	1.943	10.116	0.928	0.656	1.477
		Mp	**Mn**	**Mw**	**Mz**	**D**	**% < 10⁴**	**% > 10⁶**
		57,123	18,822	97,988	340,729	5.21	11.39	0.09
Sample-B	Autoclave LDPE	**M bulk**	**Mz+1**	**MLS/Mcc**	**MLS/Mncc**	**IV bulk**	**g-prime**	gpcBR
		423,612	753,307	3.021	21.022	0.887	0.501	3.472
		Mp	**Mn**	**Mw**	**Mz**	**D**	**% < 104**	**% > 106**
		45,322	20,150	140,227	548,847	6.96	10.61	0.83

3D-TREF Summary Report
Job : Tubular vs Autoclave LDPE

SM No. Sample ID Lot No.	Zone-1 Solubles @ 40°C			Zone-2 Polymer 40°C < Temp <110°C			trefBR=K*Mw^0.73/IV-1 K=0.000374 for PE
	Wt. %	Cal'd Mv	IV g/dl	Wt. %	Cal'd Mv	IV g/dl	Remarks
Sampe A Tubular LDPE	7.0%	5,000	0.17	93%	197,000	1.03	
					trefBR =	1.66	
Sampe B Autoclave LDPE	5.0%	14,000	0.16	95%	403,000	0.98	Autoclave LDPE has higher LCB and lower TREF elution temperature.
					trefBR =	3.72	

Figure 8.
3D-GPC-TREF summary report for tubular versus autoclave LDPE.

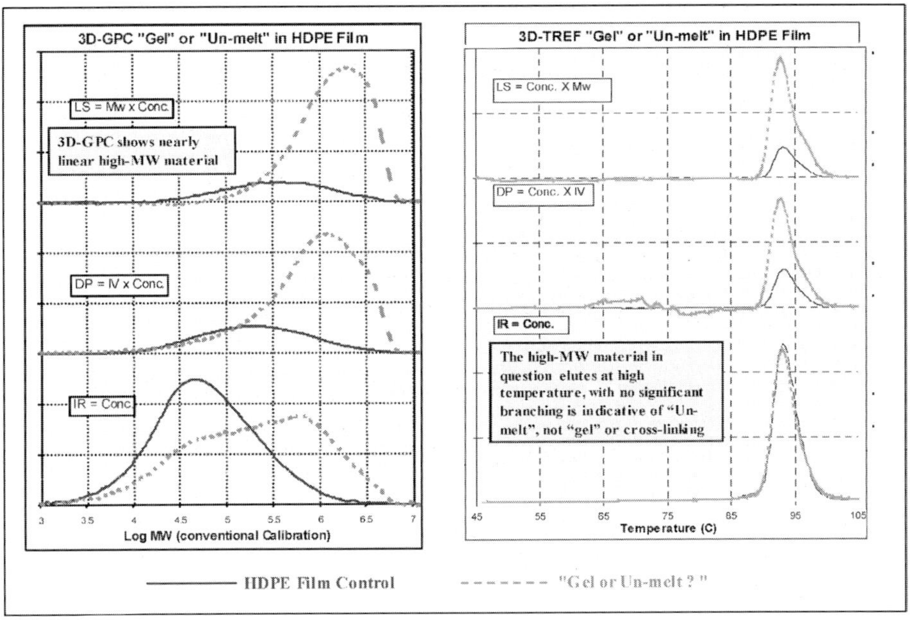

Figure 9.
3D-GPC-TREF determination of "gel or un-melt?" structure in HDPE film.

sample sizes makes it very difficult for them to be studied by most other analytical tests. This makes 3D-GPC-TREF uniquely useful for samples as little as a fraction of a milligram. Highly cross-linked gels may exist in samples that may not dissolve in hot TCB. These gel structures would not be injected in either the GPC or TREF column, and would not be detected. The presence of these gel fractions can be estimated by the percent mass recovery data in the GPC or TREF experiment.

In the next study example, we show that one can distinguish different types of LCB in polymers and that LCB is not limited to LDPE. LCB can exist in HDPE made from ZN slurry process or as a result of thermal degradation and chain extension processes in melt extrusion. While the 3D-GPC side of the μ-structure plot in Figure 10 shows the existence of LCB in all three samples, 3D-GPC is not capable of distinguishing the LCB types. However, the LCB in LDPE is clearly distinguished from that in the HDPE samples by the 3D-TREF result shown in Figure 10. The two HDPE samples with LCB content elute from TREF at high temperature, but the LDPE sample has a lower TREF elution temperature because of its high content of butyl (C4) branching. *These C4 branches are the result of the chain-end backbiting branching mechanism in the high-pressure LDPE process. It is kind of fascinating to picture that the basic structure of LDPE minus the presence of LCB would not be very different from a single-site hexene LLDPE copolymer of very narrow CCD.*

Moderation of Co-Crystallization Effect in TREF Separation

Co-Crystallization of polymer blends is a complicating factor that can compromise the accuracy of the compositional analyses by TREF and by crystallization analysis fractionation (Crystaf).[8] Strong co-crystallization distorts the elution profile of TREF and Crystaf separations. Evidence has been reported that the co-crystallization problem is less severe in TREF than in Crystaf. These reports conclude that TREF is more appropriate for analyzing samples with complex CCDs because it provides better peak resolution of blends. Nevertheless,

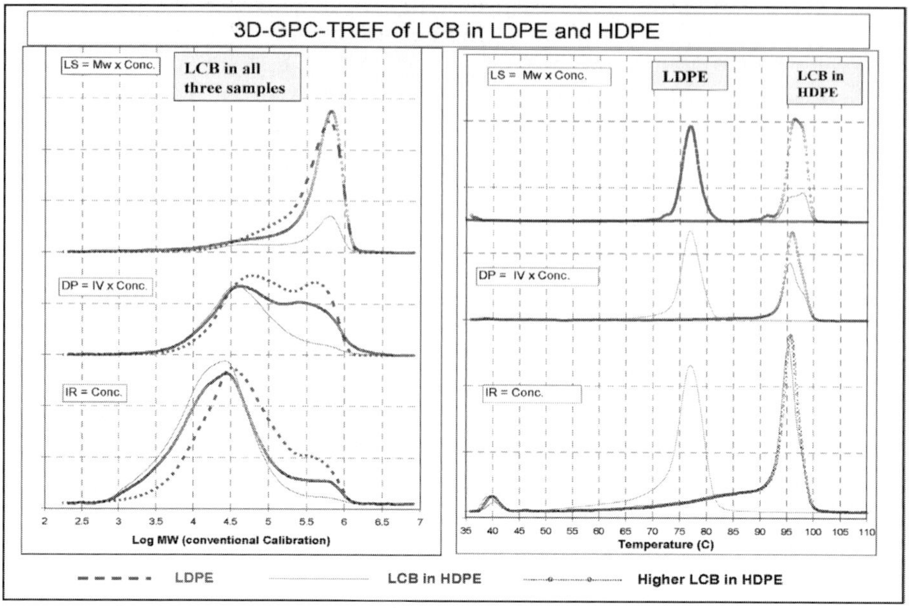

Figure 10.
LCB in LDPE versus LCB in HDPE.

co-crystallization still is a concern in fast TREF separations using fast cooling rates.

In our study reported below, we have managed to reduce the co-crystallization effect in TREF by using glass bead packing instead of polymeric di-vinyl benzene (DVB) packing. Reasonable control of the co-crystallization effect has been made possible even at fast TREF analyses at the rate of three to four hours per sample. The results of this study are explained in Figure 11 and 12 for the DVB and glass bead TREF columns, respectively. (The glass beads GL-0191 of 18–27 microns were purchased from MO-SCI Specialty Products, LLD, 4000 Enterprise Dr., Rolla,

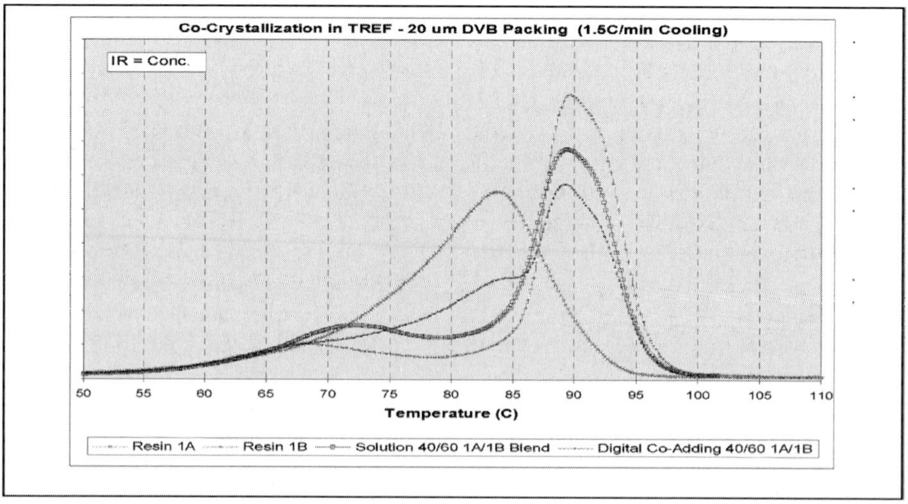

Figure 11.
Strong co-crystallization effect in a TREF column with DVB packing in a 3-hour TREF experiment.

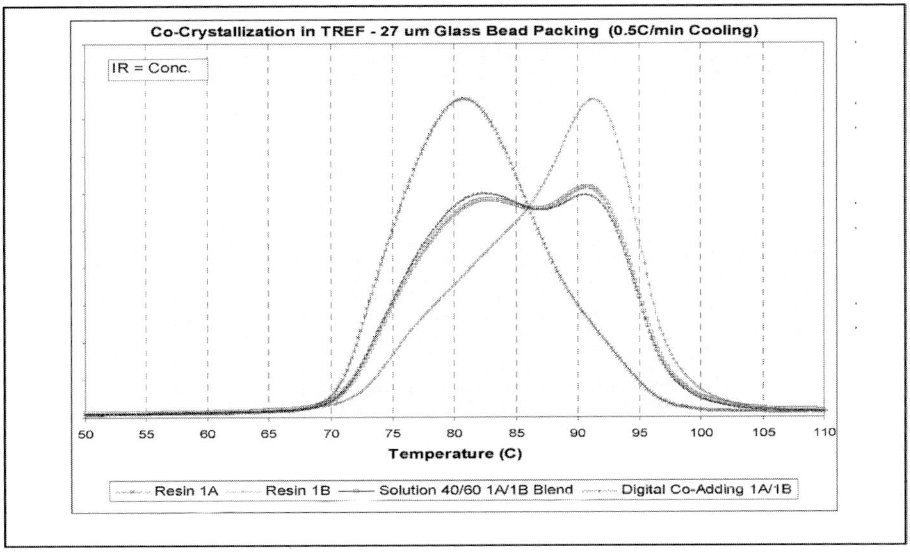

Figure 12.
Minor co-crystallization effect in a TREF column with glass bead packing in a 4-hour TREF experiment.

MO 65402.) In each figure, there are three experimental TREF elution curves: one for each of the two components in the blend and a third curve for a solution blend sample of the two components. The fourth curve in each figure is the digitally co-added results of the two component curves.

The difference between the curve of solution blend and the co-added curve provides the measure of the co-crystallization effect. It is clear that co-crystallization effect is much worse with the DVB packing in Figure 11 than for the glass bead TREF column in Figure 12. The cooling rate is found to have an effect on the degree of co-crystallization, but only to a much lesser extent than the effect caused by the difference of the TREF packing material. It is clear that, under proper TREF conditions, the co-crystallization problem can be moderated considerably as shown in Figure 12, which gives still reasonably short TREF analyses at the rate of four hours per sample.

Analyses of Resin Compositions in A Multi-Layer Polymer Film

Under the TREF conditions without strong co-crystallization, the 3D-TREF technique can be used to study the compositions of polymer blends or components in the multi-layer films. In the example study below, we were asked to find out the resin compositions in an unknown polymer film. We were provided with five resin samples that might have been used in making the film. To approach this problem, we first obtained the 3D-TREF results for the film and the five resins. We then used the Excel solver program to search for the answer. The results are explained below with the help of Figure 13 and 14 for the raw 3D-TREF data and the solver search results, respectively.

The top row in Figure 13 shows the IR, DP and LS profiles of the TREF runs for the unknown film sample. The bottom row shows the corresponding detector tracings of the TREF runs for the five resin samples. The 3D-TREF curves of the five resins are consistent with the following polymer types: 1 = an isotactic polypropylene (iPP), 2 = a metallocene LLDPE (mLLDPE), 3 = a ZN-LLDPE, 4 = a second mLLDPE, and 5 = a plastomer. The first thing we observed was that resins 2 and 4 had identical 3D-TREF profiles. They are likely the same mLLDPE product.

The goal of the study was to find the best combination of the TREF curves for these

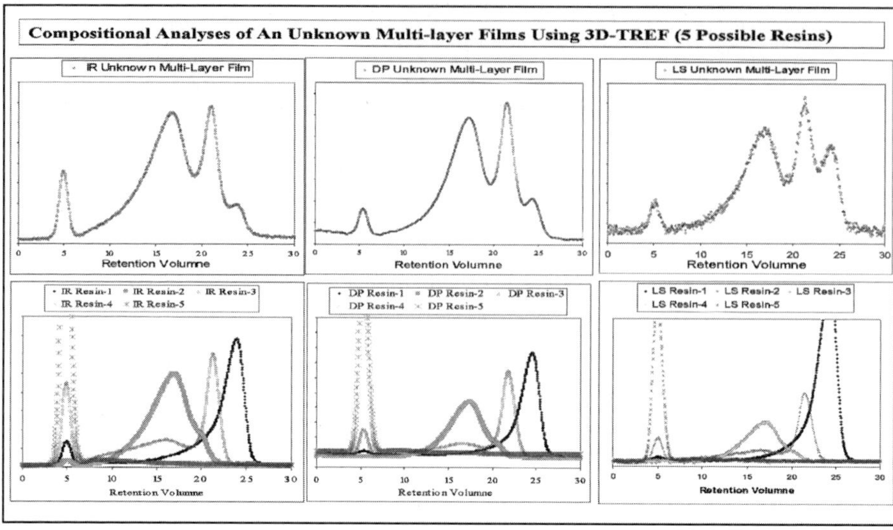

Figure 13.
Raw 3D-TREF data of an unknown multi-layer film and the five possible resin components.

resins to match those of the unknown film sample in each of the three detector signals. This was quite a curve-fitting challenge that would not have been possible without the help of a curve-fitting program like the Excel Solver. The Solver search results, shown in Figure 14, indicate that the film was made up by three resin components – iPP, mLLD, and a ZN-LLD – with the estimated weight percentage of 10, 40, and 50%, respectively. The co-added curve from this three-resin combination for each detector is shown in Figure 14 and compared with the 3D-TREF data of the

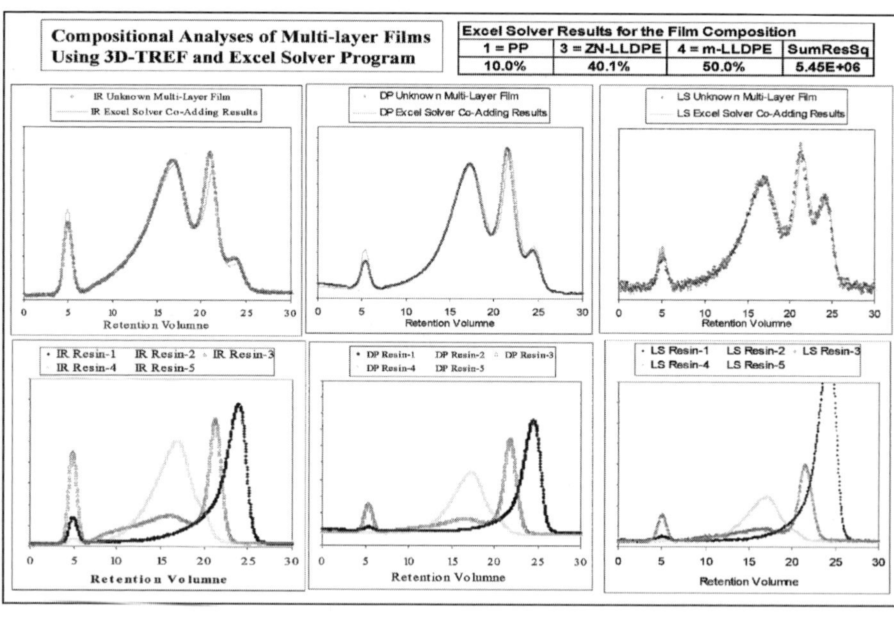

Figure 14.
Use of 3D-TREF and Excel Solver for the compositional analysis of a multi-layer film.

unknown film sample. We were pleasantly surprised to see the very good fit between the predicted and the experimental data for the film sample. With the success of this study, it is possible to say that under favorable conditions, the 3D-TREF can be quite useful for quantitative studies of polymer blends and can be used as a tool for the compositional reverse-engineering of fabricated products.

Characterization of Reverse MW-SCB Dependency

Comonomer incorporation is a very important practice for optimizing the properties of LLDPE and HDPE products. It is highly desirable to have the comonomers distributed more in the higher molecular weight fractions of the sample to achieve a greater effect of the comonomer presence to the polymer property. For catalysis research and product development, there is a strong need for analytical techniques that can determine the molecular weight dependency of comonomer and SCB distribution in polymer products. One approach has been reported with the use of an online Fourier transform IR detector (FTIR) in a GPC experiment.[9] However, in a rather roundabout way, some aspect of the polymer MW-SCB dependency can also be determined by the 3D-TREF method. This can be explained with the help of the sketches shown in Figure 15 and 16.

Figure 15 depicts a GPC experiment where a ZN-LLDPE sample is separated across the MWD in the x-direction. The downward pointed wide-arrow in the middle of the Figure is an illustration for this ZN-type MW-SCB dependency, i.e., the SCB (or co-monomer content) decreases with increasing molecular weight. What researchers wish to achieve is the reversed MW-SCB dependency in the product, i.e., the MW-SCB trend depicted by the upward pointed wide-arrow in the Figure. The nature of the MW-SCB analyses is obviously a two-dimensional problem. We need comonomer detection (e.g. FTIR) in the y-direction to complement the GPC-MWD information in the x-direction. This is how the GPC-FTIR method works. *Alternatively, one can also get the SCBD information by a cross-fraction of GPC fractions by a second TREF analysis, as depicted on the left y-axis of the sketch. This would be called a GPC-TREF cross-fractionation.*

To explain how 3D-TREF works to determine MW-SCB dependency, we turn the GPC-based sketch in Figure 15 into the TREF-based sketch in Figure 16 by a 90° rotation of the plot. In this case, we now

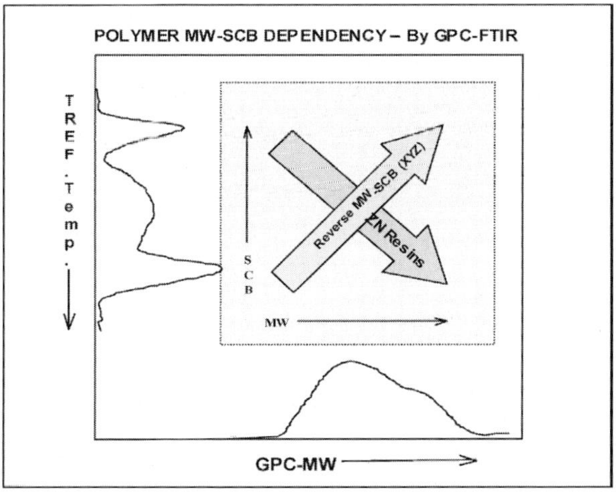

Figure 15.
A sketch to explain the determination of polymer MW-SCB dependency by GPC-FTIR analysis.

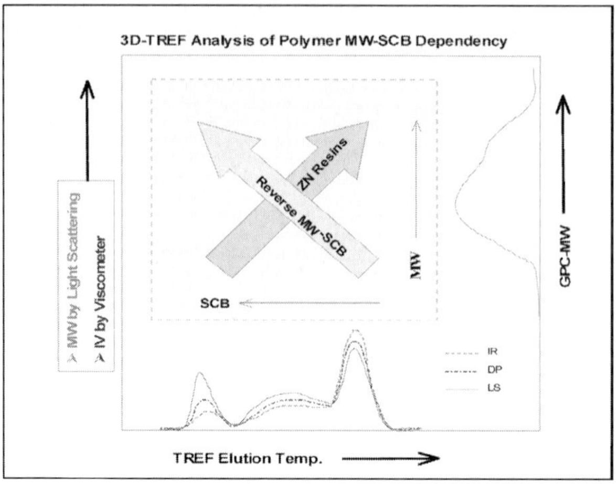

Figure 16.
A sketch to explain the determination of polymer MW-SCB dependency by 3D-TREF analysis.

have the SCB and comonomer content information displayed across the x-direction. The wide-arrows for the ZN and reversed molecular weight dependency are still in place to show the desired information we try to determine. What is needed in this case is the molecular weight information in the y-direction. This need is met obviously by the LS-M_w and Viscometer-IV information derived from the triple-detector features of the 3D-TREF technique. *If GPC is used to get the MWD information of the TREF fractions, this situation of course becomes the technique known as the TREF-GPC cross fractionation.*

An example of a typical 3D-TREF report on polymer MW-SCB dependency is shown in Figure 17 for two LLDPE samples: one is a ZN type and the other is of the reversed MW-SCB type produced with a mixed metallocene catalyst. To help the visualization of the relative molecular weight dependency across the TREF curves, the 3D-TREF curves in this figure are plotted in the so-called "relative-scale" mode, where the three detector signals are normalized to the same height at the peak of the highest IR peak. In this plot option, the relative molecular weight and IV trend in the three temperature zones can easily be recognized visually. Typically, we report the results in three zones, i.e., the "soluble fraction" zone 1, the "high-impact" zone 2, and "homopolymer fraction" zone 3. The elution temperature limits of these zones can be varied to fit the specific need of the 3D-TREF study. If needed, reporting the results in more number of temperature zones can be easily accommodated. In the example shown in Figure 17, these three zones are chosen to be the soluble fraction that elutes below the temperature of 40 °C, zone 2 of elution temperature between 40 °C and 85 °C, and zone 3 for fractions at elution temperatures higher than 85 °C.

For sample A of the ZN-LLDPE type shown at the upper left corner of Figure 17, we see first that the multiple TREF peaks reflecting the multiple sites are clearly separated. Second, we see that the LS signal decreases toward the lower-melting fractions. That means that the comonomer fraction is higher in the lower molecular weight chains in zone 2. This is a MW-SCB trend that is opposite to what a researcher would have wished to make. On the other hand, for sample B from a mixed mLLDPE catalyst, we see that LS signal is higher for the lower melting peak of higher comonomer content. This gives a clear indication that the comonomer fraction is indeed higher in the higher molecular weight

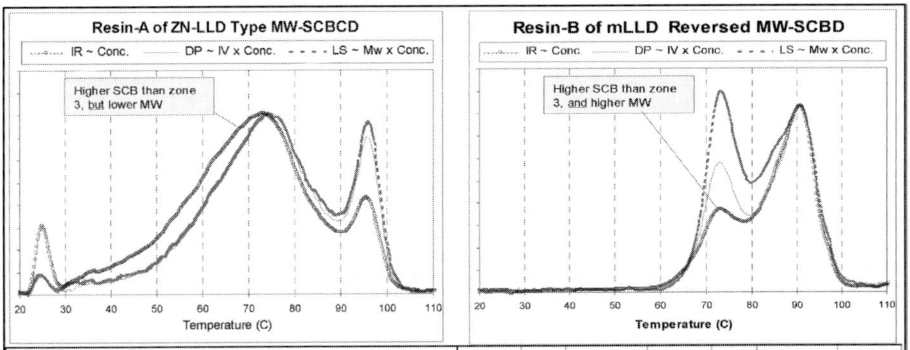

3D-TREF Study of the MW-SCB Dependency

SM No. Sample ID Lot No.	Zone-1 Solubles Temp < 40°C			Zone-2 Low-Melts 40°C < Temp <85°C			Zone-3 High-Melts 85°C < Temp < 120°C			Ratio Zone2 / Zone3		
	Wt. %	Mw	IV g/dl	Wt. %	Mw	IV g/dl	Wt. %	Mw	IV g/dl	Wt. %	Mw	IV g/dl
Resin-A ZN-LLD	13.7%	63,000	0.72	66%	107,000	1.41	20%	174,000	2.17	3.32	0.61	0.65
Resin-B mLLD Reversed MW-SCBD	0.0%	~	~	46%	108,000	1.04	54%	67,700	0.89	0.85	1.60	1.18

Figure 17.
Quantification of the reverse MW-SCB dependency of LLDPE samples.

chains in this sample, as one would have hoped to achieve in catalyst research. To quantify this reversed MW-SCB dependency, we process the TREF results in the three zones to provide the values of weight percentage, M_w, and IV in each zone. The results of such calculations are shown in the bottom table of Figure 17. The ratio of the M_w and IV values between zone 2 and zone 3 is also reported in the last column of the table. The degree of reversed MW-SCB dependency can therefore be quantified by theses ratio values. The level of reverse MW-SCB dependency is proportional to how much higher the molecular weight ratio exceeds the value of 1.0. Therefore, the reversed MW-SCB dependency in sample B is reflected in the high molecular weight ratio of 1.60, while the value is 0.61 and less than one for sample A of ZN-LLDPE type.

The molecular weight and the amount of the soluble fraction, *or the lack of it*, are important pieces of information of the samples as well, especially for studying high-impact polypropylene, thermoplastic olefin (TPO) and tie-layer products that depend on the solid state rubbery-versus-crystalline phase separation in their end-use applications. Some examples for such applications are presented below.

Structural Studies of Functional Polyolefin

The value of the sub-ambient TREF capability of our instrument using ODCB solvent is demonstrated in the study example of the two commercial tie-layer products shown in Figure 18. The rubber phase components in the two products both elute as the "soluble fractions" in the 25 °C TREF experiment shown in the top two curves. The important compositional difference of the rubber components used in the two products can only be revealed in the sub-ambient 0 °C TREF experiment shown in the two bottom curves.

Good use can be made of the dual-wavelength feature of the IR4 detector in our 3D-GPC-TREF system. One wavelength (IR1) in this detector has comparable detection sensitivity for both the aliphatic methylene (CH_2) and methyl

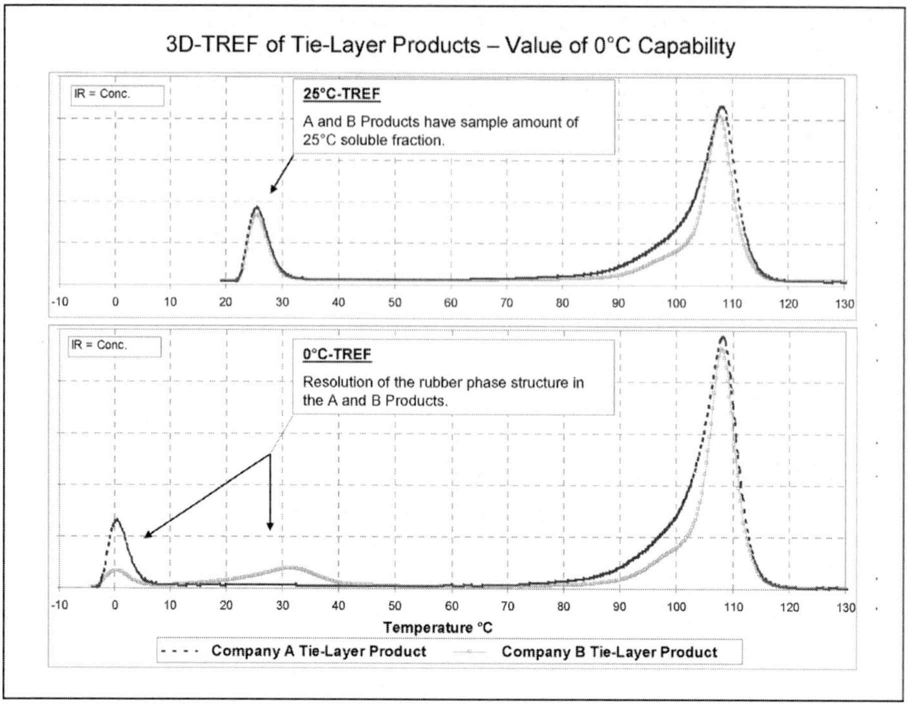

Figure 18.
Value of sub-ambient TREF capability for the compositional study of the rubber phase components in tie-layer products.

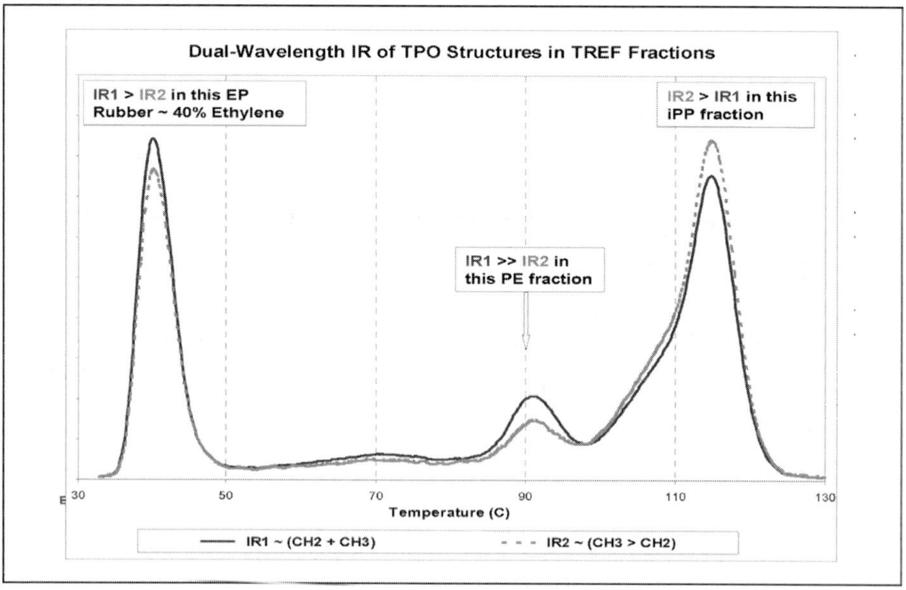

Figure 19.
Dual-wavelength IR detection of TPO structures in TREF fractions.

(CH_3) groups in polyolefins, while the 2nd wavelength (IR2) is more sensitive for detecting CH_3 than the IR1 wavelength. An example of the dual-wavelength IR application is shown in Figure 19 for a TREF study of a polypropylene-based TPO sample. With the presence of both IR1 and IR2 detector signals, one can see the ethylene-propylene compositional differences across the TREF curve by observing the relative peak heights between the IR1 and IR2 signals. By doing so, one can easily identify the structure of the three main peaks in the TREF curve as the following: a large isotactic polypropylene component at high-melting, a small HDPE peak near 90 °C and a large EP copolymer rubber component in the soluble fraction peak.

In the study example below, the effect of visbreaking on HDPE was studied with and without a functional modifier. The results are shown in Figure 20. Under the condition of no modifier, one sees only the effect of the chain scission effect of the visbreaking process. In this case, we see a decrease in molecular weight (lower LS signal) of the product as compared to the starting HDPE material. The visbroken product in this case still has the microstructure of a HDPE, as indicated by the fact that the TREF peak remains at the same high elution temperature position. For the case where a functional modifier is included in the visbreaking process, one sees that there is an additional shift of resulting material to a lower TREF elution temperature. This observation of the down shift in elution temperature is of value to learn about how the modifier molecules are incorporated into the final product. The TREF result suggests that the grafting of the modifier to polyethylene has more likely occurred onto the chain backbone, rather than to the chain ends. This is because, only side chains in the polyethylene backbone will decrease TREF elution temperatures. Modifiers added to the polyethylene chain ends would not have a strong effect on the TREF elution temperature.

In this next study of two high impact polypropylene samples shown in Figure 21, the goal of the study is to find ways of modifying the features of company A polymer to mimic that of the company B product. From the 3D-GPC results, one sees that the company B product has higher molecular weight. But, there is still the question of not knowing where one needs to make a boost in molecular weight, whether it needs be in the rubber phase component or in the crystalline component of the company A product. The answer can

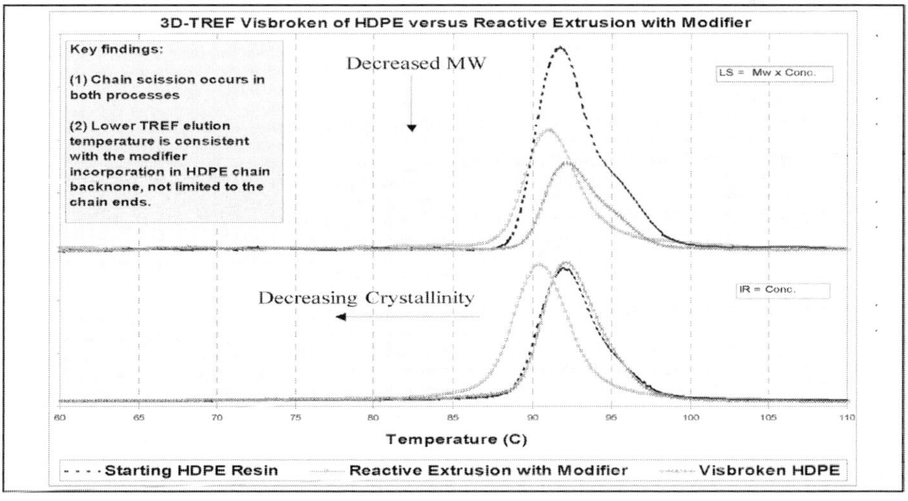

Figure 20.
3D-TREF study of HDPE visbreaking versus reactive extrusion with modifiers.

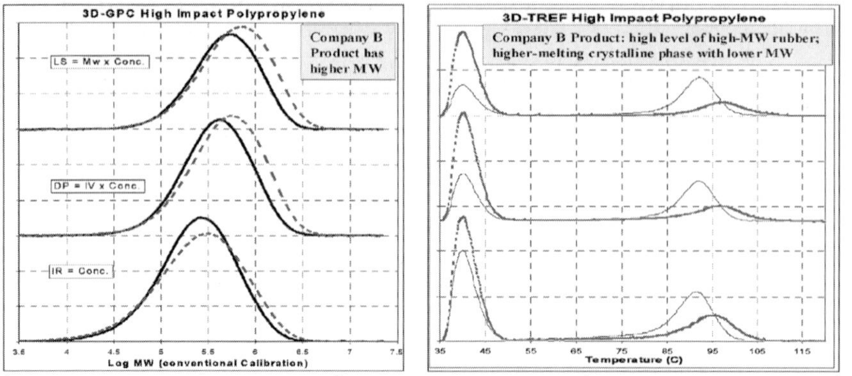

Sample ID	Soluble Fraction @40°C			Polymer Fraction @Temp>40°C		
	%	Mw (LS/IR)	IV (DP/IR)	%	Mw (LS/IR)	IV (DP/IR)
——— Company A Product	48	338,000	1.57	52	635,000	2.09
– – – Company B Product	64	523,000	2.58	36	354,000	1.52

Figure 21.
Reverse engineering of high-impact polypropylene using 3D-GPC-TREF.

be clearly seen in the 3D-TREF results of the two samples. The 3D-TREF result shows clearly that the increase in molecular weight needs to be in the rubbery component of the product. In fact, the molecular weight difference between the two products shown in the table at the bottom the Figure 21 also provides the information about how much increase of the rubber molecular weight is needed.

Concluding Remarks

We found that the combination of 3D-GPC and 3D-TREF in a form of a hybrid system is a very powerful toll for studying polyolefin μ-structures. Most of the successful applications of the technique in solving company problems are credited to a very large part to the high precision capability of our instrument. Automation of sample injection and temperature programming in our system is the key to our high precision analyses. Analytical information provided by this technique has the value to greatly shorten the product development time. Not only is 3D-GPC-TREF useful in understanding the μ-structure of a given resin, it also has been useful in analyzing the composition of polymer blends and fabricated products. In today's challenging times in polyolefin industry, a strong program in 3D-GPC-TREF technology is a highly desirable element in a modern day polyolefin research organization.

Acknowledgements: The author wishes to acknowledge the technical assistance of Tia Kuhlman, Steve Elmore and Sue Setty for setting up and carrying out most of the 3D-GPC-TREF experiments during the course of this work.

[1] Waters Corporation, 34 Maple Street, Milford, MA 01757, USA.
[2] Polymer ChAR, Valencia Technology Park, P. O. Box 176 E-46980 Paterna, Spain. E-mail Info@polymerchar.com
[3] Precision Detectors, 34 Williams Way, P. O. Box 738, Bellingham, MA 02019, USA. E-mail Lighscatter@netscape.net
[4] Polymer Laboratories, Amherst Fields Research Park, 160 Old Farm Road, Amherst, MA 01002, USA. E-mail SalesUS@polymerlabs.com
[5] W. W. Yau, D. Gillespie, *Polymer* **2001**, *42*, 8947–8958; W. W. Yau, TAPPI 2005 PLACE Conference Proceedings, TAPPI PRESS, Atlanta, Session 19, Paper 19-1.

[6] W. W. Yau, TAPPI 2005 PLACE Conference Proceedings, TAPPI PRESS, Atlanta, Session 19, Paper 19-1; C. T. Enos, K. Rufener, J. Merrick-Mack, W. Yau, Waters International GPC Symposium Proceedings, June 6–12, **2003**, Baltimore, MD USA.
[7] R. Shroff, H. Mavridis, *Macromolecules* **1999**, *31*(25), 8454–8464.
[8] S. Anantawaraskul, J. B. P. Soares, P. M. Wood-Adams, *Macromol. Symp.* **2004**, *206*, 57–68; S. Anantawaraskul, J. B. P. Soares, P. M. Wood-Adams, *Journal of Polymer Sci.: Part B: Polymer Physics* **2003**, *41*, *1762–1778*.
[9] P. J. DesLauriers, D. C. Rohlfing, E. T. Hsieh, *Polymer* **2002**, *43*, 159–170.

DOI: 10.1002/masy.200751104

Separation and Characterization of Ethylene-Propylene Copolymers by High-Temperature Gradient HPLC Coupled to FTIR Spectroscopy

Andreas Albrecht,[1] *Lars-Christian Heinz,*[1] *Dieter Lilge,*[2] *Harald Pasch**[1]

Summary: The chromatographic separation of ethylene-propylene (EP) copolymers with regard to chemical composition was accomplished by a new technique - high-temperature gradient HPLC. Using a mobile phase of ethylene glycol monobutylether (EGMBE) and 1,2,4-trichlorobenzene (TCB), and silica gel as the stationary phase, copolymers with different ethylene contents were separated according to their chemical compositions. Using a sample solvent of n-decanol and a column temperature of 140 °C, chromatographic conditions were established that correspond to separation in a precipitation-redissolution mechanism. With the aim to obtain further information on the separation process, the HPLC system was coupled to FTIR spectroscopy through a LC-Transform interface. The FTIR data confirmed that the copolymers were separated according to the ethylene content of the eluted samples.

Keywords: ethylene-propylene copolymers; FTIR; high performance liquid chromatography; polyolefins

Introduction

Ethylene-propylene (EP) copolymers are important polymeric materials. Depending on the comonomer content of the EP copolymers their properties change from crystalline (low comonomer content) to amorphous (high comonomer content). The determination of the chemical composition distribution (CCD) of EP copolymers requires fast and efficient analytical methods. The analysis of EP copolymers and blends of polyethylene and polypropylene by DSC is a well established and simple method;[1–6] however, it is not a separation method and, thus, cannot provide a CCD.

For separations according to chemical composition, temperature rising elution fractionation (TREF) is used.[7–14] In a TREF experiment the sample is dissolved at high temperature and precipitated using a slow cooling process. In a second step, fractions of the precipitated material are eluted at increasing elution temperatures. The fractionation occurs mainly regarding sample crystallinity and cannot be used for amorphous samples. In addition, the long analysis times and the complexity of the method prevent a routine use on a daily basis. With crystallization analysis fractionation (CRYSTAF), which is partially similar to TREF but requires less analysis time, determinations of the CCD of EP copolymers, high density polyethylene (HDPE), low density polyethylene (LDPE), linear low density polyethylene (LLDPE) and polypropylene (PP) were accomplished.[15–19] It should be noted that amorphous EP copolymers that do not crystallize cannot be separated using the mentioned fractionation techniques. Other techniques that separate polyolefins according to chemical composition are selective extraction with appropriate solvents and solution-precipitation techniques.[20–22] These

[1] Deutsches Kunststoff-Institut (German Institute for Polymers), Schloßgartenstr. 6, 64289 Darmstadt, Germany
E-mail: hpasch@dki.tu-darmstadt.de
[2] BASELL GmbH, Industriepark Hoechst Gebäude C 657, 65926 Frankfurt, Germany

techniques, however, require significant amounts of solvents, time and labour.

Within the large variety of liquid chromatographic techniques, only size exclusion chromatography (SEC) has been used so far for the analysis of polyolefins. High temperature SEC enables a more or less correct determination of molar mass distributions of olefin copolymers, but components with different chemical compositions may co-elute. Even by coupling SEC with FTIR spectroscopy, only average chemical compositions corresponding to each elution volume, but not CCDs, can be determined.[23–25]

High performance liquid chromatography (HPLC) is an important tool for fast separation of complex polymers with regard to chemical composition.[26,27] Different separation mechanisms such as adsorption-desorption or precipitation-redissolution are used.[28,29] Typical concentration detectors like differential refractive index (DRI) and evaporative light scattering detectors (ELSD) do not provide information on the chemical composition of the separated species. When HPLC is coupled to FTIR, however, information on the chemical composition of the chromatographic fractions can be obtained.[30–32]

Unfortunately, up to now, HPLC techniques for the separation of polymers have been used only at temperatures below 100 °C. For polyolefins, however, temperatures between 130–160 °C are necessary for keeping the polymer samples in solution. In our previous work,[33] a gradient HPLC system was developed that enables to separate blends of olefin homopolymers (PE and PP) according to their chemical composition at high temperatures. Polar silica gel, as the stationary phase, and a mobile phase comprising ethylene glycol monobutylether (EGMBE) and 1,2,4-trichlorobenzene (TCB) were used. The separation was based on the fact that EGMBE is a non-solvent for linear polyethylene (above ~20 kg/mol) and a solvent for isotactic polypropylene. Thus, polyethylene precipitates on the column while polypropylene is eluted when pure EGMBE is used as the mobile phase. With a gradient of EGMBE and TCB, the separation of polypropylene and polyethylene can be achieved, where polypropylene elutes in the SEC mode and polyethylene elutes with the solvent gradient, respectively.[33]

In the present paper, the separation of EP copolymers by high temperature gradient HPLC according to chemical composition is reported. For the first time the coupling of gradient HPLC with FTIR spectroscopy at temperatures that are suitable for the characterization of polyolefins is described.

Experimental Part

High-Temperature Chromatograph PL XT-220

A prototype high-temperature gradient HPLC system PL XT-220 (Polymer Laboratories, Varian Inc, Church Stretton, England) was used.[34] The stationary phase was silica gel Nucleosil 500, column size 25 × 0.46 cm I.D., average particle diameter 5 μm (Macherey Nagel, Düren, Germany). For dissolution and injection of the samples a robotic sample handling system PL-XTR (Polymer Laboratories) was used. The temperature of the auto sampler with the sample block and the injection needle, the injection port and the transfer line between the auto sampler and the column compartment was set to 140 °C. The mobile phase flow rate was 1 mL/min and 50 μL of sample solutions were injected. The polymers were dissolved in TCB at a concentration of 1–1.2 mg/mL at a temperature of 150 °C. The column outlet was connected either to an evaporative light scattering detector (ELSD, model PL-ELS 1000, Polymer Laboratories) or to a LC-Transform FTIR Interface (Series 300, Lab Connections, Carrboro, USA). The ELSD was operated at a nebulisation temperature of 160 °C, an evaporation temperature of 270 °C and an air flow of 1.5 mL/min. The LC-Transform was operated at a stage temperature of 164 °C and a temperature

Table 1.
Average molar masses, polydispersities (PD) and ethylene contents of the copolymers.

Sample	M_w (SEC) [g/mol]	M_n (SEC) [g/mol]	PD	Ethylene (FTIR) wt [%]
EP1	277,000	77,600	3.57	49
EP2	270,000	49,000	5.52	32
EP3	757,000	97,700	7.76	26
EP4	277,000	72,400	3.82	38

of the transfer line of 150 °C. A temperature gradient of the nozzle was used due to the different boiling points of the mobile phase components. The rotation velocity of the Germanium disc was 10 degree/min. FTIR spectra of the deposited fractions were recorded using a Nicolet Protegè 460 (Thermo Electron, Waltham, USA). For data collection and processing, the WinGPC-Software (Polymer Standards Service, Mainz, Germany) was used.

Crystaf

A Crystaf model 200 (Polymer Char S.A, Valencia, Spain) was used for the fractionation. For the analysis, 20 mg of the sample were dissolved in 30 mL of 1,2-dichlorobenzene in a stirred vessel. A cooling rate of 0.1 °C/min was used for the fractionation of the samples.

Solvents

1,2,4-Trichlorobenzene (TCB), n-decanol and ethylene glycol monobutylether (EGMBE), all of synthesis quality (Merck, Darmstadt, Germany) were used as received in this study.

Samples

Linear polyethylene (PE) with a weight-average molar mass of M_w 126 kg/mol was obtained from Polymer Standards Service (Mainz, Germany). Moplen HP 400R (M_w: 305 kg/mol) is a commercial polypropylene of BASELL Polyolefine GmbH, Frankfurt, Germany. A blend of 2.52 mg PP and 2.26 mg PE was used for developing a coupling method HT-gradient HPLC-FTIR.

EP copolymers were obtained from BASELL Polyolefine GmbH and their characteristic data are summarized in Table 1. The samples were prepared by the gas phase process.

Results and Discussion

Separation of EP copolymers

The fractionation of amorphous EP copolymers with regard to chemical composition is difficult. These samples do not crystallize and, hence, TREF or CRYSTAF cannot be used for fractionation. The CRYSTAF curves of samples EP1-EP4 are presented in Figure 1. As can be seen, there are no crystallizing fractions indicating that the samples are fully amorphous. On the other hand, the CRYSTAF results indicate that the samples do not contain any PE or PP homopolymers. These components would crystallize producing peaks in the CRYSTAF curves of Figure 1.

For the chromatographic separation of the copolymer samples regarding chemical composition, gradient HPLC is the method of choice. In gradient HPLC experiments, very frequently the sample is dissolved in a good solvent and then injected into a mobile phase of low solvent strength or even a non-solvent. This causes the sample to precipitate on the column. By stepwise or continuously increasing the solvating power of the eluent, the precipitate is re-dissolved and separated by adsorptive or solubility effects. These adsorptive or solubility effects correlate with the chemical composition of the sample and separation according to chemical composition can be achieved.

As has been shown previously, PE-PP blends can be separated by high-temperature gradient HPLC using silica gel as the stationary phase. A mobile phase of TCB as the thermodynamically good solvent and EGMBE as the poor solvent is used. TCB is a good solvent for both PE and PP, while EGMBE is a good solvent for PP and a non-solvent for PE.[21,35] Starting with 100% EGMBE for 2 min, the volume

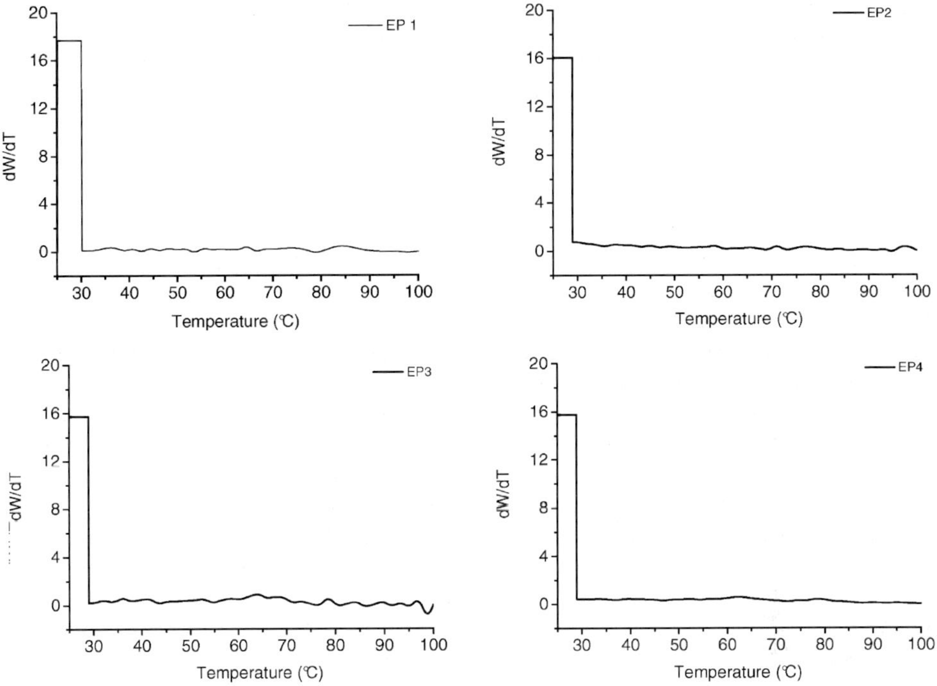

Figure 1.
CRYSTAF curves of samples EP1, EP2, EP3 and EP4.

fraction of TCB is increased linearly to 100% in 3 min and held constant for 3 min. Finally, the initial chromatographic conditions are re-established. The corresponding gradient profile is shown in Figure 2.

Because of the column void volume and the dwell volume of the chromatographic system, the gradient reaches the detector with a delay of 5.03 min, i.e., the gradient reaches the detector at 7.03 min. For the EP copolymers, chromatograms are obtained that are presented in Figure 3. Qualitatively similar chromatograms are obtained for all samples. All samples elute in three different elution regions. These can be attributed to different separation mechanisms and

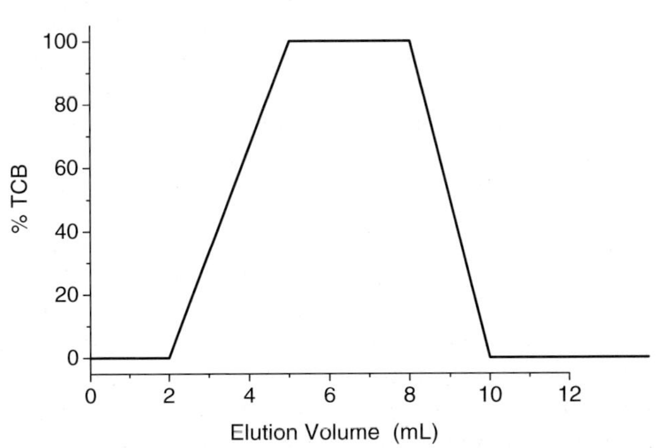

Figure 2.
Gradient profile, stationary phase: Nucleosil 500, mobile phase: linear gradient EGMBE-TCB.

different chemical compositions. The polymer fractions that appear between 1.5 and 2.2 mL elute with the initial mobile phase and, thus, are soluble in EGMBE at 140 °C. Taking into account the results of our previous investigations[33] these fractions can be assumed to be EP copolymers with low ethylene contents. In addition, these fractions could contain some atactic PP. The next elution region correlates with the solvent peak (n-decanol) at an elution volume of 2.3 mL. The polymer fractions that elute with the sample solvent n-decanol are assumed to be EP copolymers with medium ethylene contents. Polymers that are insoluble in EGMBE and do not elute with n-decanol appear later in the chromatograms. They elute with the TCB gradient in the elution volume region of 7–9 mL. These polymer fractions are supposedly EP copolymers with higher ethylene contents. As has been shown by the CRYSTAF experiments (see Figure 1) the samples do not contain PP or PE homopolymer fractions.

Coupling of HT-gradient HPLC and FTIR

As is shown in Figure 3 the chromatograms indicate that the EP copolymer samples are separated with regard to chemical composition by high-temperature gradient HPLC. To prove that this is the case and to get more information on the separation mechanism, the HPLC system was coupled to FTIR spectroscopy. The most suitable way of combining the two techniques is by using the LC-Transform interface where the eluate coming from chromatography is sprayed on a rotating Germanium disc. Under high temperature spraying conditions the mobile phase evaporates and the polymer fractions are deposited as solid layers at different positions of the disc. Subsequently, the disc is placed in a FTIR spectrometer and spectra are taken from all polymer fractions.

One major issue of the spray deposition procedure is the boiling point of the mobile phase. In particular, when conducting gradient HPLC separations, problems arise

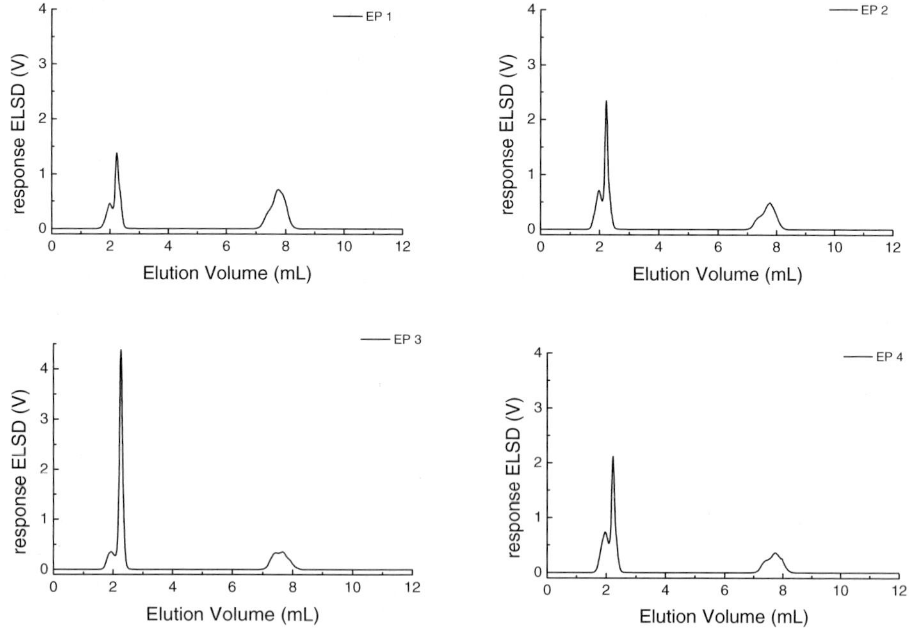

Figure 3.
Gradient HPLC chromatograms of samples EP1-EP4, stationary phase: Nucleosil 500, mobile phase: linear gradient EGMBE-TCB, column temperature: 140 °C, detector: ELSD, sample solvent: n-decanol, sample concentration ~ 1.3 mg/mL.

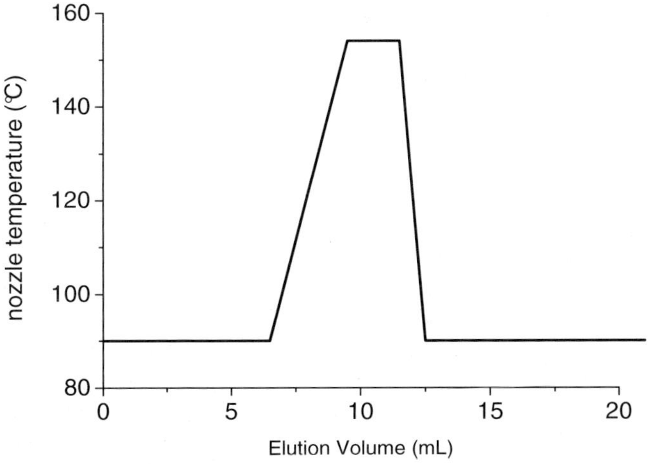

Figure 4.
Temperature gradient for the spray nozzle of the LC-Transform interface.

from the fact that the components of the mobile phase have different boiling points. In the present case these are 220 °C for TCB and 170 °C for EGMBE. Therefore, the evaporation temperature in the LC-Transform interface has to be changed continuously and corresponding to the actual mobile phase composition. A linear temperature gradient for the LC-Transform nozzle has been found to be the optimum, see Figure 4. Starting with a nozzle temperature of 90 °C for 6.5 min, the nozzle temperature is raised linearly to 151 °C in the following 3 min and than kept constant for another 3 min. Finally, the nozzle temperature is decreased back to 90 °C.

To prove that the system is working properly, a model PP-PE blend was fractionated and analyzed by HPLC-FTIR. The concentration profile of this separation obtained by the ELS detector is shown in Figure 5.A, indicating that a perfect separation into two components was obtained. The first peak eluting at 2 mL corresponds to PP and the later eluting peak corresponds to PE. The Gram-Schmidt plot resulting from summarizing all FTIR peak intensities in the range of 2800–3200 cm^{-1} as a presentation of the concentration profile is given in Figure 5.B. The comparison of the Gram-Schmidt plot (Figure 5.B) with the concentration chromatogram

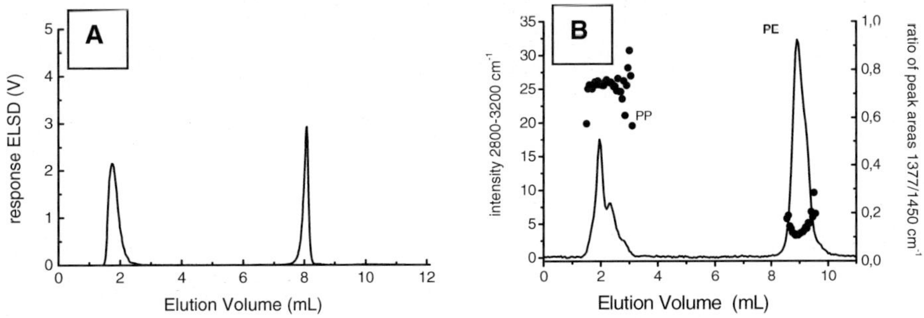

Figure 5.
HT-HPLC separation of a PP-PE blend (A) and HPLC/FTIR analysis (B), full line ELSD trace (A) or Gram-Schmidt plot (B), ■ relative amount of CH_3 groups (peak ratio CH_3/CH_2), chromatographic conditions see Figure 3.

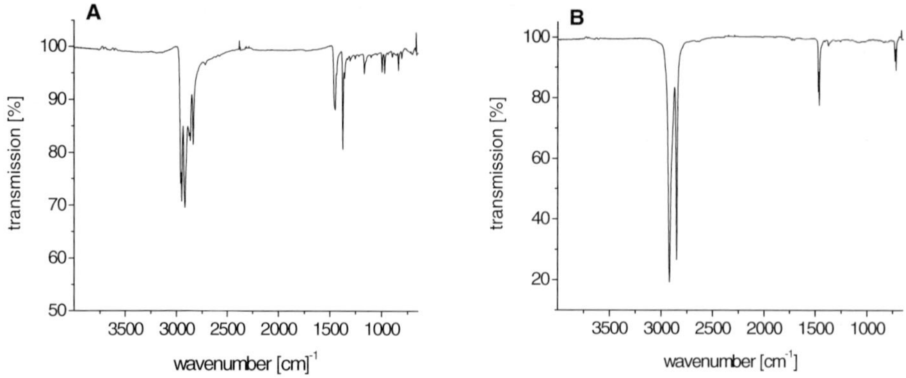

Figure 6.
FTIR spectra of the PP and PE peaks (in Figure 5) taken at peak maximum.

(Figure 5.A) indicates that the first eluting peak appears to be broader in the Gram-Schmidt plot. An explanation for this behaviour is the use of n-decanol as the sample solvent (boiling point 233 °C) that cannot evaporate as fast as EGMBE at the present conditions. Traces of n-decanol that were not evaporated were detected even on the Germanium disc.

The quantitative analysis of the FTIR spectra was carried out by analyzing the absorption bands for the CH_3 groups at 1376 cm^{-1} and the CH_2 groups at 1462 cm^{-1}.[36] The presence of CH_3 groups indicates branching and is characteristic for propylene units while the concentration of CH_2 groups is a measure for total polymer concentration. In Figure 5.B the ratio of CH_3/CH_2 (1394–1328 cm^{-1}/1394–1488 cm^{-1}) is plotted against the elution volume.

For the PP-PE blend in Figure 5, peak area ratios of 0.75 for the PP peak and 0.10 for the PE peak were detected. The low concentration of CH_3 groups in the PE fraction indicates that the material is slightly branched. The corresponding FTIR spectra at the peak maximum verify the chemical composition of the first peak as PP and of the second peak as PE, respectively, see Figure 6.

Analysis of EP copolymers by HPLC-FTIR

As has been demonstrated, PP-PE blends can be separated by HT-gradient HPLC with regard to chemical composition. The quantitative analysis of the fractions is conducted by FTIR spectroscopy using the LC-Transform approach. A similar procedure shall be used also for the analysis of the EP copolymers. Using the same experimental conditions for the LC-Transform and the FTIR measurements, the EP copolymers are analysed with regard to chemical composition as is shown in Figure 7.

A comparison of the Gram-Schmidt plots and the elution profiles obtained with the ELS detector shows that the Gram-Schmidt peaks are slightly broader. This is due to the fact that a certain spreading cannot be avoided in the spraying procedure of the LC-Transform interface. This, however, does not affect the quality of separation with regard to chemical composition.

The peak area ratios CH_3/CH_2 for the different elution peaks show clearly that early eluting peaks exhibit a higher peak area ratio than late eluting peaks. This clearly indicates that the early eluting fractions have lower ethylene contents than the late eluting fractions. Over the entire elution volume range the ratio of CH_3/CH_2 decreases with an increase in the elution volume. This is in full agreement with the proposed separation mechanism.

If the ethylene content in the copolymer increases, a decrease in solubility is observed. Copolymers with high propylene content (peak area ratio >0.55) elute in EGMBE or

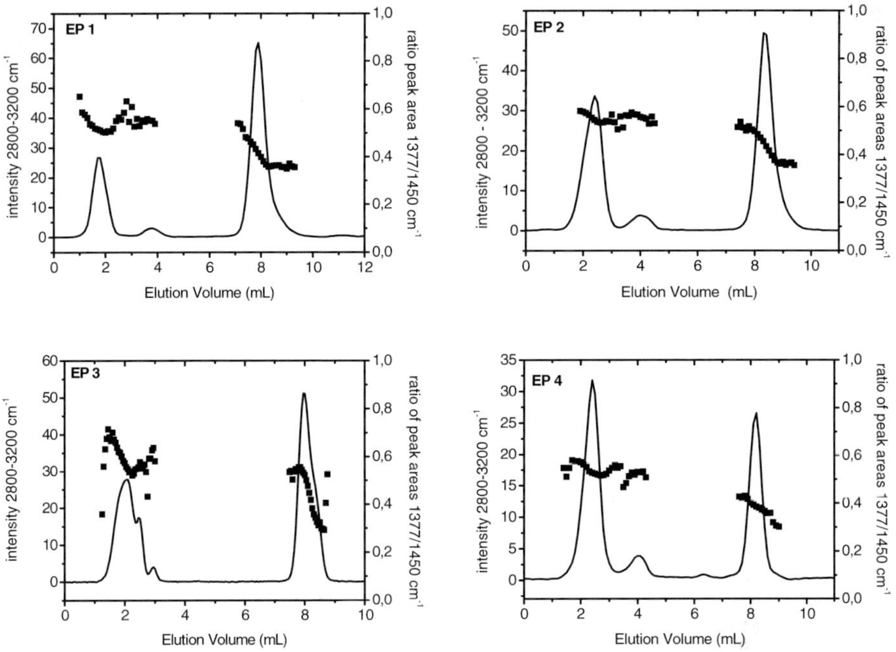

Figure 7.
Gram-Schmidt plots of the HT-gradient HPLC separation of EP copolymers 1–4, full line Gram-Schmidt plots, ■ relative amount of CH_3 groups (peak ratio CH_3/CH_2), chromatographic conditions see Figure 3.

with the sample solvent peak. From the fact that the peak area ratio is less than for pure PP (0.75 in Figure 6) it can be concluded that the early eluting peaks do not contain PP but EP copolymer with low ethylene contents. This is in a good agreement with the CRYSTAF measurements, where no isotactic PP and no PE were detected. With increasing PE content the EP copolymer becomes less and less soluble in EGMBE and elutes with the solvent gradient according to the solvent strength. The EP copolymers start to elute close to the starting point of the gradient between 7.1 and 7.5 mL. Apparently, there is a solubility threshold at a certain copolymer composition. Accordingly, rather similar CH_3/CH_2 ratios are obtained for the end of the first peak and the beginning of the second peak. For the EP copolymer fractions with the lowest propylene content, peak area ratios of 0.25 to 0.35 are measured. Compared to the blend separation in Figure 5 a shift to lower elution volumes is observed for the late eluting peaks. This additionally indicates that the peaks are due to EP copolymer and not PE homopolymer. Finally, the decreasing peak area ratio with increasing elution volume within the late eluting peaks clearly confirms the separation with regard to chemical composition. Due to the steep gradient it is not surprising that the peak positions of the EP copolymers do not vary much. For a better separation of the fractions A more shallow gradient has to be chosen.

As mentioned earlier, the higher boiling point of n-decanol compared to EGMBE results in a certain spreading of n-decanol on the Germanium disc. The elution peak appearing at an elution volume of 4 mL corresponds to the solvent peak in Figure 3. This peak contains small amounts of EP copolymer with medium ethylene content as can be seen in Figure 7.

As indicated in Figure 7, the analyzed EP copolymers exhibit broad distributions with chemical compositions ranging from 0.7 to 0.3 (relative amount of CH_3/CH_2). The differences between the samples can be understood from the chemical compositions of the different eluting fractions. Sample EP1 has the highest average ethylene content (49%) while sample EP3 has the lowest average ethylene content (26%), see Table 1. The results in Figure 7 indicate the same trend: in EP1 the first eluting (propylene-rich) fraction shows a peak area ratio of 0.63–0.5 while for EP3 this ratio is higher with 0.72–0.5. This indicates that the propylene content of this fraction is higher for sample EP3. In addition to the higher propylene content, the area of this peak is higher as compared to EP1. This is a clear proof that the average propylene content of EP3 is higher as compared to EP1.

The aim of forthcoming investigations is the exact calibration of the FTIR system for a strict quantitative analysis of the ethylene content of the EP copolymers. To achieve this goal NMR data of model EP copolymers will be correlated with corresponding FTIR bands for PP and PE. By using a fully calibrated system, absolute chemical composition values of amorphous EP copolymers can be obtained and the chemical composition distribution of these copolymers can be described.

Conclusions

High temperature gradient HPLC is the first chromatographic method that enables separation of EP copolymers with regard to their chemical composition. Using a polar stationary phase and a solvent gradient of EGMBE/TCB it is possible to fractionate complex copolymers in the direction of increasing ethylene content. The separation of a PE-PP blend and EP copolymers into components of different chemical compositions was confirmed by coupling of the gradient HPLC with FTIR. This is the first time that a LC-Transform FTIR interface was used for gradient-HPLC at temperatures suitable for the separation of polyolefins.

Acknowledgements: The authors would like to acknowledge the excellent collaboration with A. Williams, S. O'Donohue and R. Linz (Polymer Laboratories, Church Stretton, England) in designing, building and adjusting the high-temperature gradient HPLC system. We thank BASELL Polyolefine GmbH for supplying the EP copolymers. The financial support of the work by BASELL Polyolefine GmbH (Frankfurt, Germany) and BMBF (grant No. 03C035YA) is highly appreciated.

[1] J. Weng, R. H. Olley, D. C. Bassett, P. Jääskeläinen, *J. Polymer Sci. B Polymer Phys.* **2004**, *42*, 3318.
[2] A. K. Gupta, S. K. Rana, B. L. Deopura, *J. Appl. Polym. Sci.* **1992**, *44*, 719.
[3] Z. Bartczak, M. Pracella, *European Polymer Journal*, **2006**, *42*, 1819.
[4] S. Wang, D. Yang, *Polymer*, **2004**, *45*, 7711.
[5] V. Mathot, T. Pijpers, W. Bunge, *Polym. Mater. Sci. Eng.* **1992**, *67*, 143.
[6] D. M. Sarzotti, J. B. P. Soares, L. C. Simon, L. J. D. Britto, *Polymer*, **2004**, *45*, 4787.
[7] L. Wild, *Adv. Polym. Sci.* **1990**, *98*, 1.
[8] L. Wild, *Trends Polym. Sci.* **1993**, *1*, 50.
[9] J. B. P. Soares, A. E. Hamiliec, *Polymer*, **1995**, *36*, 1639.
[10] S. A. Karoglanian, I. R. Harrison, *Polym. Eng. Sci.* **1996**, *36*, 731.
[11] J. J. Mara, K. P. Menard, *Acta Polym.*, **1994**, *45*, 378.
[12] P. L. Joskowicz, A. Munoz, J. Barrera, A. J. Müller, *Macromol. Chem. Phys.* **1995**, *196*, 385.
[13] G. W. Harding, A. J. van Reenen, *Macromol. Chem. Phys.* **2006**, *207*, 1680.
[14] N. Aust, M. Gahleitner, K. Reichelt, B. Raninger, *Polymer Testing*, **2006**, *25*, 896.
[15] B. Monrabal, *Macromol. Symp.* **1996**, *81*, 110.
[16] B. Monrabal, *New trends in polyolefin science and technology*, In: S. Hosada, Ed., Research signpost, **1996**.
[17] B. Monrabal, J. Blanco, J. Nieto, J. B. P. Soares, *J. Polym. Sci. Part A Polym. Chem.* **1999**, *37*, 89.
[18] R. Brüll, V. Grumel, H. Pasch, H. G. Raubenheimer, R. Sanderson, U. M. Wahner, *Macromol. Symp.* **2002**, *81*, 178.
[19] H. Pasch, R. Brüll, U. Wahner, B. Monrabal, *Macromol. Mater. Eng.* **2000**, *46*, 279.
[20] H. Dexheimer, O. Fuchs, *Makromol. Chem.* **1968**, *96*, 172.
[21] A. Lehtinen, R. Paukkeri, *Macromol. Chem. Phys.* **1994**, *195*, 1539.
[22] A. Barbalata, T. Bohossian, G. Delmas, *J. Appl. Polym. Sci.* **1992**, *46*, 411.
[23] L. Verdurmen-Noël, L. Baldo, S. Bremmers, *Polymer*, **2001**, *42*, 5523.
[24] A. Faldi, J. B. P. Soares, *Polymer*, **2001**, *42*, 3057.

[25] C. C. Tso, P. J. DesLauriers, *Polymer*, **2004**, *45*, 2657.
[26] G. Glöckner, *Gradient HPLC of Copolymers and Chromatographic Cross Fractionation*, **1991**, Berlin, Springer-Verlag.
[27] H. Pasch, B. Trathnigg, *HPLC of Polymers*, **1997**, Berlin, Springer-Verlag.
[28] M. A. Quarry, M. A. Stadalius, T. H. Mourey, L. R. Snyder, *J. Chromatogr. A*, **1986**, *358*, 1.
[29] M. A. Stadalius, M. A. Quarry, T. H. Mourey, L. R. Snyder, *J. Chromatogr. A*, **1986**, *358*, 17.
[30] J. Adrian, E. Esser, G. Hellmann, H. Pasch, *Polymer*, **2000**, *41*, 2439.
[31] S. J. Kok, C. A. Wold, Th. Hankemeier, P. J. Schoenmakers, *J. Chromatogr. A*, **2003**, *1017*, 83.
[32] L.-C. Heinz, A. Siewing, H. Pasch, *e-polymers*, **2003**, 065.
[33] L.-C. Heinz, H. Pasch, *Polymer*, **2005**, *46*, 12040.
[34] L.-C. Heinz, T. Macko, A. Williams, S. O'Donohue, H. Pasch, *The Column (electronic journal)*, **2006**, *2*, 13–19.
[35] T. Macko, H. Pasch, Y. V. Kazakevich, A. Y. Fadeev, *J. Chromatogr. A*, **2003**, *988*, 69.
[36] J. E. Marks, *Polymer Data Handbook*, **1999**, Oxford University Press Inc.

 DOI: 10.1002/masy.200751105

Molecular Topology Fractionation of Polystyrene Stars and Long Chain Branched Polyethylene Fractions

David M. Meunier,[*1] *Theodore M. Stokich Jr.,*[1] *David Gillespie,*[2] *Patrick B. Smith*[1]

Summary: Control of long chain branching (LCB) architecture is an area of considerable interest in materials science because LCB can have a dominating effect on polymer rheology and properties. Currently no analytical technique provides a quantitative description of the LCB topologies in these materials beyond a basic estimation of the average number of branch points per molecule. Neither the molecular weight of the branch, nor the shape of the branched molecule (e. g. star, comb, "H" or other) can be determined using current state of the art methodology such as size exclusion chromatography (SEC) with molecular weight sensitive detectors or nuclear magnetic resonance spectroscopy.

In our laboratory, we have developed a fractionation method that sorts polymer solutes based on LCB topology. The approach, which we term molecular topology fractionation (MTF), utilizes a separating medium comprising channels having dimensions similar in size to the dimensions of the macromolecules being analyzed. An applied flow field provides the driving force for the separation. Although the details of the separation mechanism are not well understood at this time, two possible mechanisms are being considered. In one, dissolved solute molecules are restricted by the channels such that the relaxation modes for reorientation determine the rate of transport. In the second, pinning (or entanglement) of molecules on the stationary phase determines the rate of transport. Both mechanisms result in the largest molecules eluting latest (opposite to the sequencing in SEC), and produce significant additional retardation for LCB chains above that of linear chains. This additional retardation leads to fractionation of an LCB distribution even if the hydrodynamic radii of the components are the same.

In this paper, an overview of the MTF experiment will be provided. MTF fractionation of PS stars is presented to demonstrate the separation of LCB chains from linear chains and LCB chains based on topology. The application of MTF for characterizing LCB polyolefin fractions will be shown. The paper will also include a brief discussion of the coupling of MTF and SEC in an on line two dimensional approach for determination of LCB distributions.

Keywords: branching distribution; characterization; entanglement; lightly cross-linked homogeneous ethylene octene copolymer; long chain branching; pinning; star polystyrene; topology fractionation

Introduction

Molecular topology fractionation[1] (MTF) is a relatively new dilute solution polymer separation technique developed within the Dow Chemical Company.[2] The development

[1] The Dow Chemical Company, Analytical Sciences, Midland, MI
E-mail: DMeunier@dow.com
[2] The Dow Chemical Company, Polyolefins Research, Freeport, TX

of MTF was driven by a need for better tools for characterizing long chain branching (LCB). Present state-of-the-art techniques such as 13C NMR[3] and size-exclusion chromatography (SEC) with molecular weight sensitive detectors[4] provide information about the average number of branches per molecule. Neither technique provides information about the molecular weights of the branches nor the shapes of the branched molecules. Additionally, for low levels of branching, both techniques are operating at relatively low signal to noise, making detection of low levels of LCB a challenge.

Long chain branching is introduced into polymers because it can have substantial impact on the rheological behavior of the system provided the branch lengths are significantly larger than the entanglement molecular weight.[5] Because their radii of gyration are smaller, LCB polymers shear thin to a greater extent than linear polymers of the same molecular weight. On the other end of the shear spectrum, LCB polymers offer higher zero shear viscosity than their linear counterparts. This is because LCB polymers entangle much more effectively than their linear counterparts. The enhanced entanglement of LCB polymers was the motivation for the present embodiment of MTF. The idea was to create an entangling environment within a chromatographic column through which LCB molecules and linear molecules would be forced to flow. It should be noted here that the term "entanglement" in the MTF experiment is interchangeable with the term "pinning". The original idea behind MTF column development was to create a series of posts on which LCB chains or linear chains could become pinned.

In the first successful demonstration of MTF, poly(styrene-co-divinylbenzene) monolithic columns, having macropores (channels) of average diameter on the order of 100–200 nm, were used in the separation.[1] In that work, MTF was characterized by a flow rate dependent reversal in elution order of linear PS molecular weight standards. Additionally, chains possessing LCB were shown to be retained longer than linear chains of the same hydrodynamic size.

It was recognized that a second mechanism may also be operative in the MTF separation. This second mechanism involves chain restriction followed by relaxation/reorientation. In this mechanism, dissolved solutes may become restricted by a fraction of the column macropores such that the relaxation times for reorientation determine the rate of transport. Either mechanism, pinning or relaxation/reorientation, is expected to be sensitive to topology; both may be operative in a real system.

In this work, new MTF columns are introduced. The columns were prepared via high pressure packing of sub-micron, non-porous silica (surface functionalized with PS) into stainless steel columns. The new columns were used to study the elution behavior of regular PS stars. In addition, the columns were used to perform high temperature MTF on relatively narrow fractions of lightly cross-linked homogeneous ethylene octene copolymers.

Experimental Part

Materials and Samples

The silica used to pack MTF columns was obtained from Admatechs Co. Ltd. (Aichi, Japan). The product identification number was SO-C2 lot BMI206. An SEM image of these particles is shown in Figure 1. In addition, particle size distribution data provided by the vendor revealed that the average particle diameter was 0.5 micrometers and the half-width of the distribution was ~0.3 micrometers.

Narrow molecular weight polystyrene (PS) standards were obtained from Polymer Laboratories (Amherst, MA) and were used as received. Polymer standard solutions for MTF studies were prepared individually with concentrations ranging from 0.5 to 1 mg/ml depending on the molecular weight, with lower molecular weight standards prepared at higher concentration and vice-versa. The tetrahydrofuran (THF) used for standard dissolution

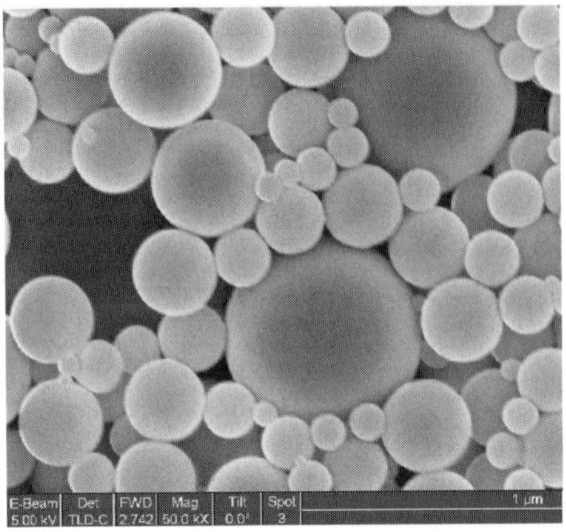

Figure 1.
SEM Image of non-functionalized Admatechs Silica. The silica was functionalized with PS and packed into a column by Mel Cabey of the Diazem Corporation.

was obtained directly from the LC reservoir so as to minimize the contributions from solvent mismatch peaks. Standard solutions were stored in the dark at ambient temperature when not in use. For SEC, standards were prepared as cocktails and the concentration of each standard in a cocktail was 0.5 mg/mL.

A three arm star PS sample was obtained from Polymer Source, Inc. (Dorval, Quebec, Canada). The star was made via anionic polymerization followed by coupling of the living anions with a trifunctional coupling agent as depicted in Figure 2. The vendor reported a weight-average molecular weight (M_W) of 1480 kg/mol for the arms of the star and 4760 kg/mol for the whole star sample. The reported whole polymer M_W does not make sense considering the sample was supposed to be a three arm star with some uniarm material present. The higher than theoretically possible M_W for the whole sample may suggest the presence of some high molecular weight impurities. Further, our characterization of the 3-arm star sample (see Results and Discussion Section) indicates that the uniarm material is actually substantially lower in molecular weight than the value supplied by the vendor.

Lightly cross-linked homogeneous ethylene octene copolymer samples were fractionated via a solvent non-solvent method to produce relatively narrow fractions for MTF. A preparative fractionation unit (PolymerChar, Valencia, Spain) was used to separate each whole polymer sample into narrow fractions with the resulting M_W's nominally from 20 to a maximum of 200–400 Kg/mole and polydispersities PDI's ranging from 1.3 to 1.5. For the whole polymer samples, short chain branch frequencies ranged from 17 to 32 short branches per thousand backbone carbon atoms, and short chain branching had minimal influence on MTF elution. Long chain branching frequencies (LCBf) of the unfractionated materials were determined by triple detector SEC[4] and ranged from

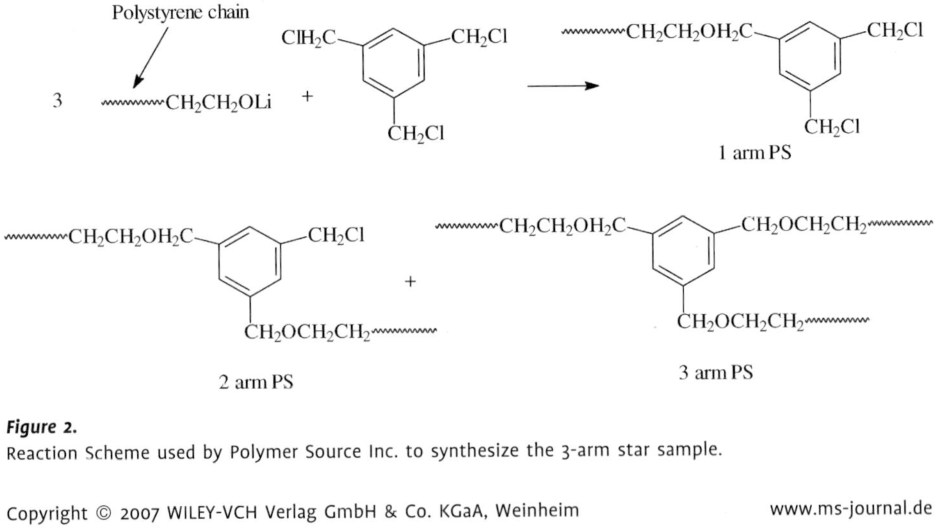

Figure 2.
Reaction Scheme used by Polymer Source Inc. to synthesize the 3-arm star sample.

0.005 to 0.7 long branches per thousand backbone carbon atoms. LCBfs were not determined for the fractions themselves. Because a fractionation was carried out, it is not clear that the average LCBfs still apply to each fraction in a series. It is expected that fractions from the LCBf = 0.005 sample would have a *very* small population of long chain branched molecules, while fractions from the LCBf = 0.70 sample would be expected to have a large population of multiply-branched LCB molecules, especially in the higher molecular weight fractions.

Equipment for MTF

The LC system used for MTF and SEC analyses consisted of a Waters Alliance 2695 pump/autosampler and a Viscotek model 302 multidetector array. The 2695 pump was used to deliver precise flowrates in the 0.01 to 0.05 mL/minute regime for MTF. HPLC grade THF (Omnisolv, HPLC Grade, EMD Chemicals) was used as the solvent and eluent. The eluent was continuously degassed. Injection volumes were 0.01 mL. The multi-detector array included UV absorbance, low angle (LALLS) and right angle laser light scattering, differential refractive index (DRI) and differential viscometer detectors. The detectors and columns were held at 50 °C for SEC. Column dimensions for MTF were 4.6 mm I. D. X 150 mm L and were packed with PS functionalized silica (made from SO-C2 described above).

A Waters Alliance Model GPCV2000 was used for high temperature MTF experiments. A flow of 0.01 ml/min was employed, and like the Waters 2695 pump used in ambient temperature experiments, the pump in the high temperature SEC was able to deliver this low flow rate both reliably and reproducibly. The carousel, injector and column regions were all controlled at 145 °C. Two detectors were used for the sample analysis: (1) A Polymer Char Model IR4 (Valencia, Spain) infrared concentration detector and (2) A Precision Detectors, Inc. (Franklin, MA) dual angle light scattering detector, Model 2040 PDI, with sensors at 15° and 90°. The pairing of these detectors provided the capability to observe the molecular weight distribution and produce an estimate of radius gyration across the elution profile of a sample.

The TCB eluent was continuously degassed by sparging the reservoir with helium and by the operation of the in-line system degasser. Injection volumes for these experiments were 0.01 mL. For high temperature MTF, concentrations were chosen to keep the product of intrinsic viscosity and concentration ($C \times [\eta]$) less than ~0.15. Samples were prepared directly in the autosampler vial by weighing an aliquot of the polyethylene fraction, adding the appropriate amount of solvent and shaking for 2 hours at 160 °C.

MTF Columns and Functionalized Packing Material

The polydisperse silica obtained from Admatechs was functionalized with PS by Mel Cabey of the Diazem Corporation (Midland, MI). Columns for MTF were packed with this PS functionalized silica by Diazem. The packing procedure consisted of slurrying the particles in a proprietary mixture of solvents and pressure packing them into the column at 6,000 PSI. Pressure was held at 6,000 PSI for 8 hours. The column outlet frit was constructed as follows. A nominal 0.5 micrometer frit was placed in the column outlet. Next, a short (e. g. a few millimeters) layer of 2 micrometer non-porous silica was packed next to the outlet frit. Finally, the column was filled with the packing particles. The column was then flushed overnight with THF or TCB before MTF experiments commenced.

SEC Experiments

SEC was carried out using the same liquid chromatography hardware as was used for MTF in order to define the polydispersity of the 3-arm star test sample as precisely as possible. A set of four mixed-B columns from Polymer Laboratories (Amherst, MA) was chosen for the separation. A flowrate of 1.0 ml/minute and an injection volume of

0.05 ml were used. Both conventional SEC results (relative to linear PS) and absolute molecular weight results (from LALLS) were determined. Data were reduced using OmniSEC software from Viscotek.

The cross-linked polyethylene whole polymers and fractions were characterized by triple detector high temperature SEC. The system consisted of a PL-220 with light scattering (PDI 2040 as described above), differential viscometry (Viscotek 4 capillary bridge design), and concentration detectors (IR4 and DRI). The column set consisted of 3 mixed-B columns from Polymer Laboratories. Sample concentrations were 1 to 2 mg/ml and injection volume was 0.100 ml. Data were reduced using custom written software.

Results and Discussion

SEC Characterization of the 3-Arm Star Sample

SEC and SEC-LALLS-DV were used to determine the molecular weight distribution (MWD) of the 3-arm star in order to compliment and aid in the analysis of MTF results. The linear PS equivalent MWD is shown in Figure 3. It is readily apparent that there are at least three components in this sample, even though they are not well resolved. Based upon calibration with linear polystyrene standards, these three components have apparent peak molecular weights of approximately 1,250, 2,500, and 3,500 kg/mole, as annotated on the distribution. It is arguable that there is a fourth component at apparent molecular weight above 3,500 kg/mole because the shape of the distribution in that regime appears slightly non-Gaussian. But if there were material eluting in that region, its molecular weight would not be known from this analysis.

The linear PS equivalent M_W and number-average molecular weight (M_N) of the 3-arm star determined from the distribution shown in Figure 3 were, 3,260 and 2,700 kg/mol, respectively. The M_W value from LALLS was, 3,760 kg/mol. The fact that the absolute M_W was larger than the linear equivalent M_W is consistent with the presence of branched molecules in the distribution. The Mark – Houwink plot from SEC-LALLS-DV, shown in Figure 4, also provides strong evidence for the presence of branched species in the distribution.

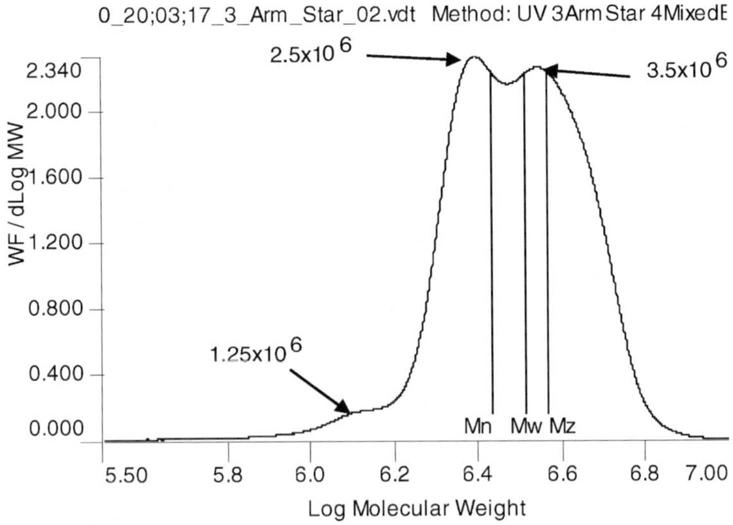

Figure 3.
Linear PS apparent molecular weight distribution of the 3-arm star sample as determined by SEC. The numbers next to the arrows represent the linear PS equivalent molecular weight at the apices of the individual peaks.

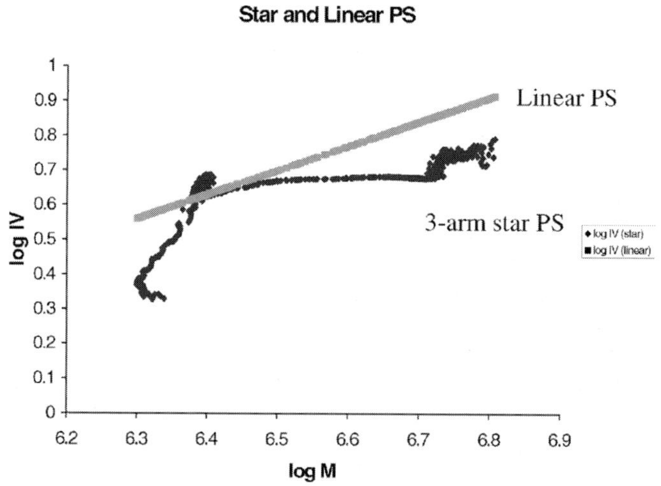

Figure 4.
Mark – Houwink plots of linear PS and star PS.

In summary, SEC characterization of the 3-arm PS star indicated that it was not a pure component, but rather consisted of a mixture of single arm, two-arm, and 3-arm species. Secondly, the molecular weights were lower than reported by the vendor for both the arms and the whole polymer sample. Arm molecular weights of ~1,250 kg/mol were found in this study, versus 1,480 kg/mol reported by the vendor. A whole polymer M_W value of 3,760 kg/mol was found here, versus 4,760 kg/mol reported by the vendor. Considering the distribution shown in Figure 3 (~5% uniarm, ~45% two arm and assuming the balance of material is 3-arm), it is not possible to rationalize the M_W value determined in this study, nor by the vendor. The only way to rationalize either reported M_W is to consider the presence of higher molecular weight species in the sample (e. g. 4-arm stars or higher).

MTF Separation of the 3-Arm Star Sample

The 3-arm star sample was injected onto an MTF column packed with the PS functionalized silica. The flowrate was 10 μl/minute, thus providing the MTF elution order reversal for molecular weights exceeding the critical molecular weight.[1] The resulting MTF fractogram, as detected by 90° light scattering is shown in Figure 5. Because the detector only responds to polymeric species, the fractogram reveals the presence of a least 4 different polymeric modes in the sample. Based on the SEC data obtained for this sample, one would have expected at least three modes in the distribution, but there was no clear evidence of a fourth mode in the SEC data.

Interpretation of the MTF fractogram of the 3-arm star sample presented a challenge. However, the combination of LALLS detection and UV detection enabled estimates of the M_W values across the MTF profile. Additionally, the MTF elution times of the species present in the 3-arm star were compared to those of linear PS standards. The ratio of LALLS detector response to the concentration detector response increased progressively from mode 1 to mode 4. In fact, the progression of the ratio followed a 1×, 2×, 3× and 4× order upon moving from mode 1 to mode 4, respectively. Thus, the elution order observed in Figure 5 is consistent with that expected for MTF[1], and the progression of molecular weight is consistent with that expected for a star sample constructed via coupling of discreet arms. However, although the 4th mode may have been anticipated from the SEC and LALLS data,

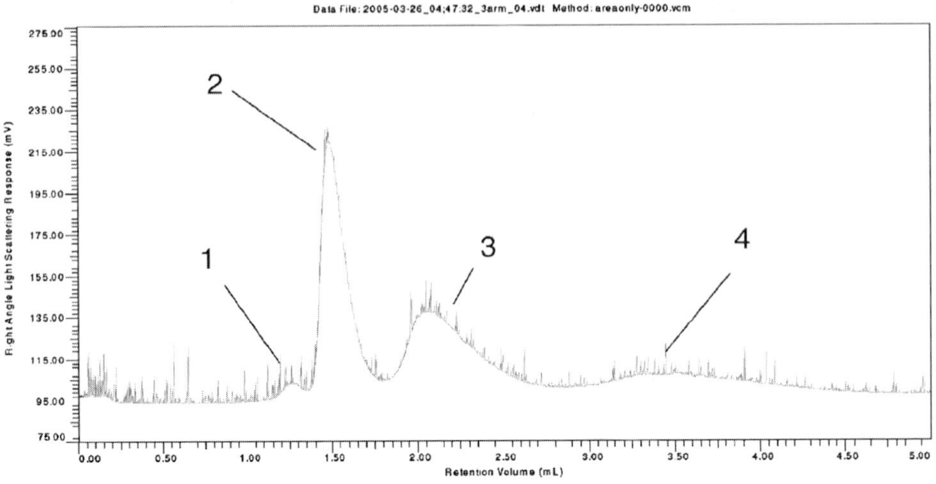

Figure 5.
MTF fractogram of 3-arm star sample as detected by 90° light scattering. The assignments of the four modes observed in the fractogram are discussed in the text. The modes are numbered 1 through 4 with 1 being the earliest eluter and 4 being the latest eluter.

its presence is now clearly revealed in the MTF experiment. Formation of the 4-arm star is discussed below.

Considering the SEC data, it would appear that a small amount of uniarm material and a relatively large amount of two-arm material was present in the sample. Both uniarm and two-arm material are linear species, and as such, their MTF elution behavior should match that of linear PS standards. A comparison of the relative peak elution volumes (or times) of linear PS standards and the peaks of the 4 modes in the 3-arm star sample is shown in Figure 6. The molecular weights of the first two modes were taken from the SEC data and assigned based on the known elution order of MTF. As can be observed, the elution volumes of the first two modes are consistent with those of linear PS standards, and thus, these two modes appear to represent the uniarm and 2-arm species present in the sample. It is clear that the latter two modes were retained far longer than the corresponding linear chains and, the LALLS/UV ratios suggest that the molecular weights of these modes are 3 and 4 times the molecular weight of mode 1, respectively.

The light scattering detector used in these studies employed detection angles of 7 degrees and 90 degrees. Although it is possible to estimate the radius of gyration (R_G) from two angle light scattering data, the expected R_G values for a multi-arm species having arm molecular weights of 1,250 kg/mol exceed the values that can be determined reliably from two angle data.[6] Calculations of the 7 degree particle scattering form factor, P(7), for PS stars[7] having arm molecular weights of 1,250 kg/mol suggest that M_W values obtained from 7 degree light scattering data should provide reasonably accurate molecular weights of the potential species present in the fractogram.

To improve the signal to noise ratio for both the concentration detector response and the light scattering response, the injected mass of the 3-arm star sample was increased by a factor of five over the standard conditions. The improved signal to noise enabled a reasonable estimate of the absolute M_W values across the MTF elution profile of modes 3 and 4. This analysis revealed that the third mode included molecular weights ranging from ~3,100 kg/mol to 3,900 kg/mol. A 3-arm

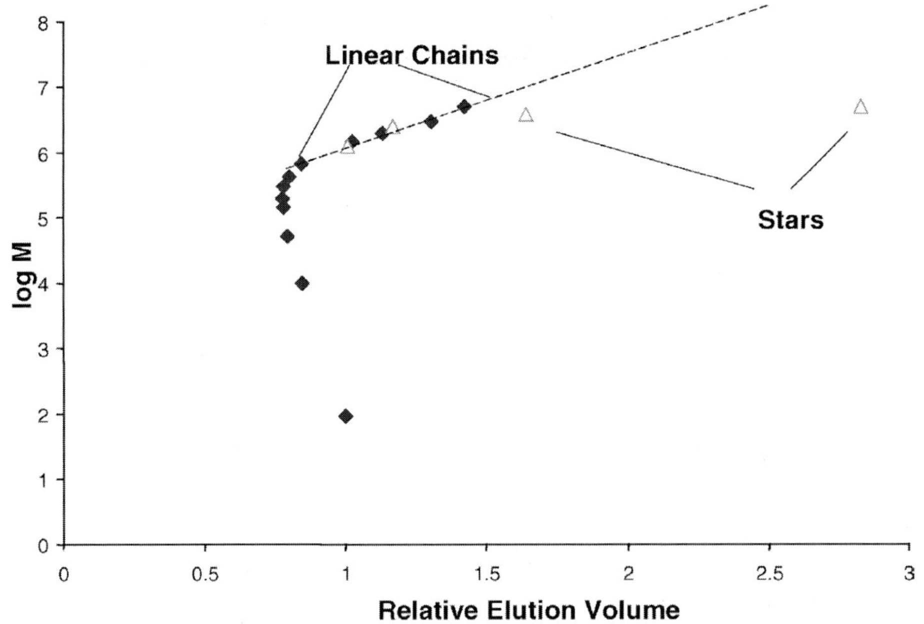

Figure 6.
Relative peak elution volume (or peak elution time) of linear PS standards (solid diamonds) and the four modes observed in the PS star fractogram (open triangles) shown in Figure 5.

star would be expected to have a molecular weight of ~3,750 kg/mol based on the molecular weight of the uniarm species. Thus, the third mode appears to represent the 3-arm star material. Finally, the fourth mode includes molecular weights ranging from ~4,500 kg/mol to ~7,000 kg/mol. Near the apex of the fourth mode, the M_W is ~5,000 kg/mol. The expected Mw of a 4-arm star is 5,000 kg/mol. The fourth mode would appear to be, predominantly, 4-arm star material. The higher molecular weights detected near the tail of the distribution most likely stem from low signal to noise in both the concentration detector and light scattering detector traces due to the very low concentration of material present.

Based on the reaction scheme used to make the star sample, one would not immediately expect the formation of a 4-arm star. However, in the synthesis of star polymers, it has been observed that molecules having more arms than the functionality of the coupling agent are often produced. The chemistry responsible for this observation is known as the lithium-halide exchange reaction.[8] In the case of the chemistry used to make this 3-arm star sample, synthesis of species having more than 3 arms is entirely possible.

Interpretation of the MTF fractogram has revealed some interesting characteristics of the MTF separation mode. First, long chain branched molecules are retained substantially longer than linear molecules of the same molecular weight. Referring to Figure 6, one can see that a 4-arm star is retained nearly a factor of two longer than a linear molecule of the same molecular weight (the highest molecular weight linear standard in the figure is approximately the same molecular weight as the 4-arm star). Additionally, the 3-arm stars are also retained much longer than linear molecules of the same molecular weight by about a factor of 1.4. Secondly, the 3-arm star and 4-arm star co-eluted in SEC, but these species are clearly separated in the MTF mode. Thus, MTF would not only appear to separate long chain branched molecules from linear molecules, but appears to separate long chain branched molecules having similar hydrodynamic volumes and differing long chain branching topology.

Sample recovery from the MTF column was also studied. For linear chains, quantitative recoveries were obtained over the molecular weight range of 10 kg/mol to 5,000 kg/mol. For the 3-arm star sample, recoveries on the order of 60% were typical. Because recoveries were quantitative for linear chains, it is assumed that the lack of recovery was due to incomplete elution of the star branched molecules. Even though recovery was below 100%, it is still clear that MTF separates based on topology. The estimate of peak M_W came from ratioing the light scattering detector signal to the concentration detector signal. The accuracy of the estimate, and hence the interpretation of the fractogram, does not require complete elution of the sample. Recent data acquired at the University of Amsterdam[9] with a similar column, revealed near quantitative elution (recovery >95%) of the same 3-arm star sample studied here. Longer run times were required to achieve the near quantitative recovery. Run time has no influence on MTF resolution. Flow rate and macropore dimensions within the column are the two largest factors influencing resolution. The fact that longer run times were required to achieve near quantitative recovery suggest that even higher order stars (more than 4 arms) were present in the sample, and more time would be needed to elute these.

MTF of Polyethylene Fractions

The molecular weight range of the polyethylene fractions relative to the MTF reversal molecular weight is an important consideration. Figure 7, for example, shows an overlay of all the MTF peak elution times for the complete set of polyethylene fractions (full range of M_W's and LCBf's), superimposed with those of the linear polystyrene standards. For the linear polystyrene standards, the M_W's have been plotted after scaling them down by a factor of 0.43 in order to adjust for the difference in the hydrodynamic volume relative to linear polyethylene. After accounting for hydrodynamic size differences, the superpositions are good as was observed previously for the PS, polybutadiene pair.[1] However, the M_W's for most fractions lie too close to the reversal or critical molecular weight, M_C, for it to be differentiated based on branching (all points in the set lie on one curve). For each LCBf, at least one fraction lies above M_C, by about a factor of two, as M_C for polyethylene would appear to be on the order of ~100 Kg/mole. Even these fractions cannot be distinguished by their peak elution times.

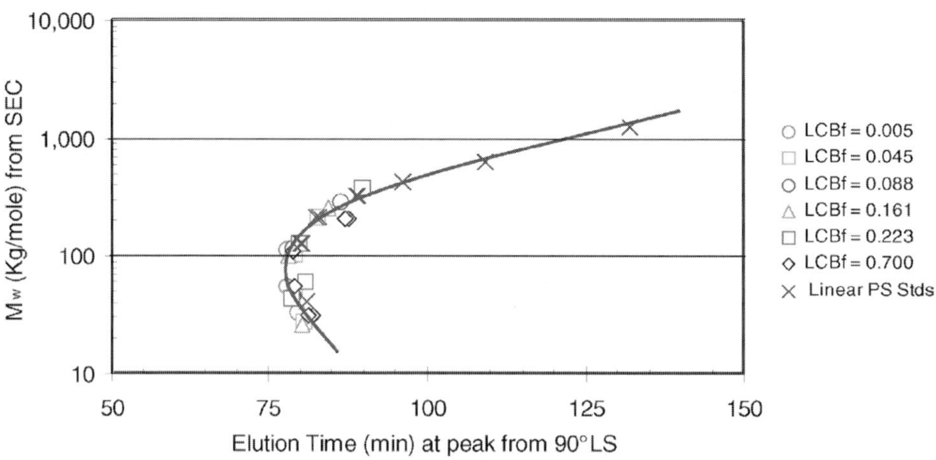

Figure 7.
MTF reversal curve comparing peak elution times of linear PS standards with those from lightly cross-linked homogeneous ethylene octene copolymer fractions.

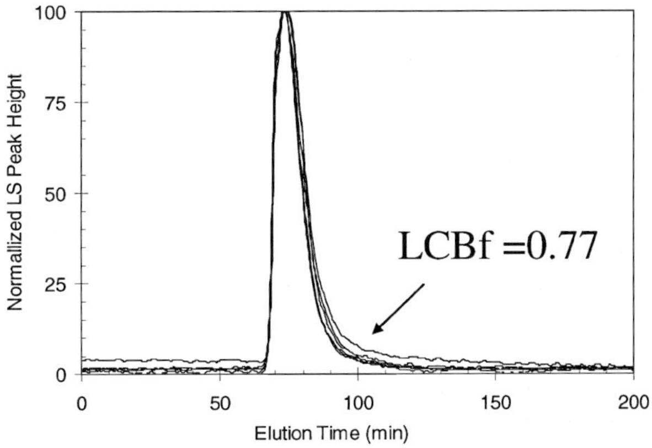

Figure 8.
MTF Fractograms of lightly cross-linked homogeneous ethylene octene copolymer fractions. All fractions had M_W values near M_C. LCBf of parent polymers from which these fractions were acquired varied from 0.005 to 0.77 per 1000 backbone carbons. Details concerning the fractions can be found in Table 1. Detection was by laser light scattering at a detection angle of 15°.

In previous work,[1] long chain branched samples had to exceed M_C before any additional retention was observed. The same observation was made in this study. Figure 8 provides an overlay of the MTF fractograms for polyolefin fractions having M_W values of ~100 kg/mol and spanning the entire range of LCBf (i. e. virtually at the M_C for this column). With the exception of a slight amount of tailing observed for the fraction coming from the highest LCBf polymer, the fractograms in Figure 8 are virtually identical. There are a couple of reasons for this. First, the fractions are too close to M_C and too close in M_W (see Table 1) to be differentiated by their peak elution times, and second, except for the highest LCBf fraction, Mark – Houwink plots from triple detector SEC indicated there was little LCB detected in the parent polymer samples near 100 kg/mol.

MTF fractograms of the highest M_W polyethylene fractions are compared in Figure 9. Details concerning the molecular weights and polydispersities of these fractions and the LCBf's of the parent polymers are provided in Table 1. Although the M_W values from SEC differ slightly among the fractions, the polydispersities, and hence the breadths of the size distributions, are all very similar. Additionally, the M_W values for these fractions all exceed M_C by more than a factor of two. The main differentiating feature among these fractions is the long chain branch frequency. For the fraction coming from the sample having the lowest

Table 1.
SEC and LCB data for polyethylene fractions.

LCBf (branches/1000 C)[a)]	M_W (kg/mol) from SEC[b)]	Polydispersity (M_W/M_N)[b)]
0.005	220/115	1.4–1.4
0.045	215/119	1.5–1.3
0.088	229/103	1.5–1.4
0.161	254/102	1.5–1.3
0.223	383/127	1.7–1.5
0.7	209/109	1.5–1.4

a) Values from parent polymer from which fractions were acquired.
b) Values from SEC based on linear HDPE, first value represents M_w of highest molecular weight fraction, while second value represents M_w of fraction closest to M_C.

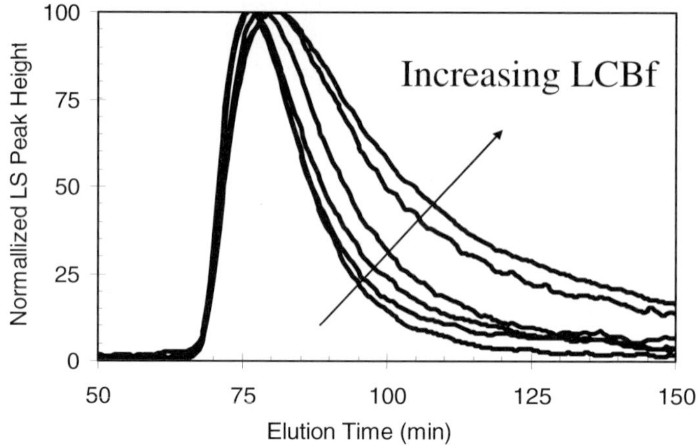

Figure 9.

MTF Fractograms of lightly cross-linked homogeneous ethylene octene copolymer fractions. All fractions had M_W values nearly two times M_C. LCBf of parent polymers from which these fractions were acquired varied from 0.005 to 0.7 per 1000 backbone carbons. Details concerning the fractions can be found in Table 1. Detection was by laser light scattering at a detection angle of 15°.

LCBf, about one chain in ten contains a long chain branch site. At the other extreme, for the fraction coming from the sample having the highest LCBf, there are about 10 branch sites per molecule.

As is evident in Figure 9, MTF is able to fractionate these higher molecular weight samples. MTF provides six distinct distributions for these samples and the main structural parameter that distinguishes these samples is the LCBf. Therefore, when there is a LCB distribution within the sample whose M_W exceeds M_C, MTF can fractionate it.

The fractionation of the polyolefin samples by MTF yields fractograms that look different from those observed for the polystyrene star. However, considering the types of branching present in the stars versus the PE fractions, one would expect the fractograms to look different. In the case of the PS star, the M_W values of the 3-arm and 4-arm material exceeded M_C, by more than a factor of 10. Additionally, the PS stars consisted of nearly discreet components, as they were made via anionic polymerization followed by coupling. On the other hand, these PE fractions come from randomly branched polymers. Thus, in each fraction, there is a distribution in number of branches and lengths of branches.

The other feature to note in Figure 9 is that not all species in these fractions appear to be MTF-effective. This is evident because the peak in the distribution is not sensitive to the LCBf. This is most likely due to the fact that a significant portion of the material in these fractions consists of linear chains. However, the breadth of the fractogram and degree of tailing both increase with increasing levels of LCB. Yet to be determined is whether the "MTF-effective" branches (i. e. those having the longest retention times) are also those that have a dominating effect on zero shear viscosity.

Finally, the use of MTF fractograms to derive a long chain branching distribution for a given hydrodynamic volume slice (e. g. from SEC) can be visualized with the aid of Figure 10. In this figure, an overlay of all three detector responses is provided for the highest LCBf fraction studied. The traces in the figure are actually the fitted results from exponentially modified Gaussian fits of the experimental data.[10] The agreement between the fitted data and experimental data was excellent. Figure 10 provides additional insight into the MTF separation. First, it

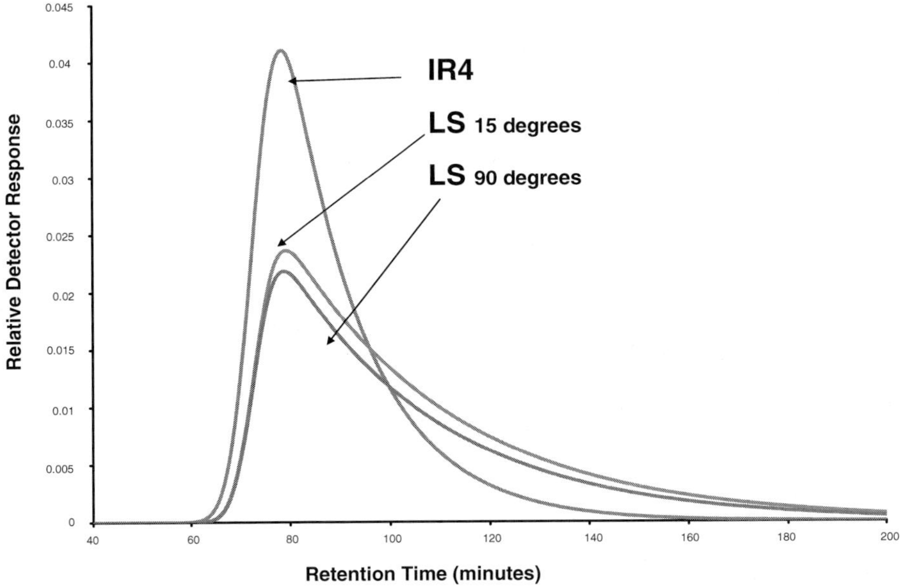

Figure 10.
Exponentially modified Gaussian fits of experimental MTF fractograms obtained for the most highly branched homogeneous ethylene octene copolymer fraction. Shown are the concentration detector (IR4), 15° and 90° light scattering traces.

can be noted that the ratio of 15° light scattering response to 90° remains roughly constant across the fractogram. This means that the R_G's of the chains eluting across the fractogram are roughly constant. Roughly constant R_G across the distribution is not surprising considering the fraction had a relatively narrow size distribution (i. e. SEC M_W/M_N ~1.4). Secondly, the ratio of 15° light scattering to IR4 (concentration) is not constant across the fractogram, but rather, it increases with increasing retention times. Thus, in MTF, chains having the same R_G's but different molecular weights are being separated, and retention increases with increasing molecular weight. At constant R_G, increasing molecular weight stems from increasing levels of branching. Thus, in MTF, chains containing the highest levels of branching are retained the longest.

In SEC, molecules are separated by hydrodynamic size, and for a sample containing LCB, this means branched chains and linear chains of the same hydrodynamic size co-elute. Among these co-eluting species, chains containing more branching will be of higher molecular weight than those containing little or no branching. In Figure 10, it was shown that MTF separates chains of the same size according to molecular weight, which means MTF separates chains according to the amount of branching present. Thus, the coupling of SEC and MTF with light scattering detection (two or more angles) would enable the determination of a branching distribution for each SEC fraction. The approach would proceed as follows: Across the MTF elution profile from each SEC fraction, both M_W and R_G are obtained from light scattering. For many linear polymers, including polyethylene, the relationship between R_G and M_W has been determined from multi-angle light scattering.[11] Alternatively, one can use the Flory-Fox equation to estimate the linear chain radii of gyration, $R_{G,L}$ for a given M_W.[12] The radii of gyration of the branched species, $R_{G,B}$, are determined from the angular dependence of their light scattering as these species elute from the MTF column. Thus, with knowledge of both $R_{G,B}$ and $R_{G,L}$, one can determine the branching index, g, across the entire MTF elution profile. The Zimm-Stockmayer model can then be used to predict the number of branch sites per chain across the MTF elution profile from g.[13] This process can be repeated for each size fraction from SEC. The distribution information potentially gleaned from the SEC-MTF-light scattering experiment cannot be obtained by any other technique at present. For example, SEC with molecular weight sensitive detectors provides only an average number of branch sites at each hydrodynamic size interval.

Particle Size Dependence of M_C

All MTF results presented to this point were obtained using columns packed with functionalized, polydisperse silica having an average particle diameter of 0.5 μm. To learn more about the particle size dependence of MTF, additional columns were packed with 0.81 and 0.75 μm silica. These silicas, obtained from Bangs Laboratories (Fishers, IN), had very narrow size distributions in contrast to the 0.5 μm material. The nominal sizes of these materials were 0.97 and 0.8 μm, respectively. The sizes reported above were estimated from SEM images. These columns were also packed by Mel Cabey, Diazem Corporation (Midland, MI) and the silica was not functionalized prior to packing.

The dependence of M_C on packing particle size is shown in Figure 11. It is clear that the critical molecular weight falls off monotonically with decreasing particle size. However, the correlation does not appear to be linear, but with only three data points, it is difficult to estimate the functional form of the relationship with any certainty. The decrease in M_C with decreasing particle size is consistent with previous MTF work performed using monolithic columns.[1] While the exact details of the MTF separation mechanism are not understood, it is clear that there is a certain critical solute size that has to be reached before the MTF mechanism is operative.

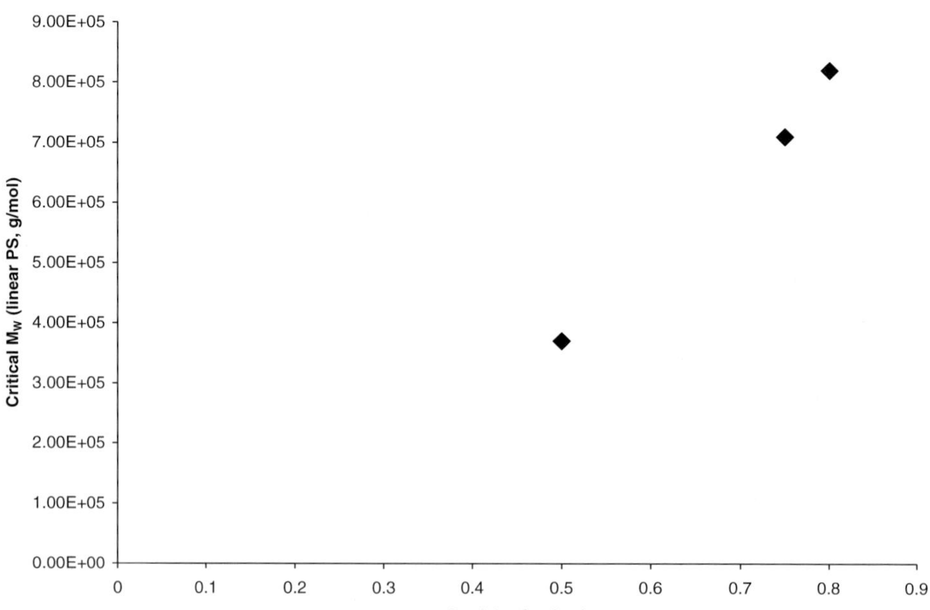

Figure 11.
Dependence of critical molecular weight (M_C) on packing particle size.

This critical size, expressed here as M_C (linear PS), decreases with particle size (i. e. macropore channel dimension). The functional form of the relationship between M_C and particle size may offer insight regarding the separation mechanism, but further work is needed to establish the relationship between M_C and particle size.

Conclusions

In a detailed analysis of a 3-arm star PS sample, it was shown that MTF can separate LCB chains from linear chains, and can separate LCB chains of differing topologies. The example shown in this paper included near baseline MTF resolution of at least 4 components present in a nominal 3-arm star sample. Despite having the same hydrodynamic volume, the 3-arm star was separated from the 4-arm star by MTF. Multidetector SEC results, knowledge of the synthesis chemistry and on-line low angle laser light scattering detection was used to interpret the MTF fractogram of the PS star sample. The clean MTF separation of star components synthesized via coupling of discreet arms will enable experimental validation of the Zimm-Stockmayer model for predicting the sizes of branched polymer chains.[13]

In a second application, polyolefin samples having similar hydrodynamic size distributions (i. e. similar breadth in SEC chromatogram) were fractionated differently by MTF with the major differentiating feature being the LCBf of the parent polyolefin polymer. For fractions exceeding M_C, MTF was shown to produce increasingly broader fractograms with increasing branching frequency.

Further interpretation of the polyolefin MTF fractograms with two angle light scattering detection revealed little change in R_G across the fractogram, but significant increases in M_W across the fractogram, especially for the highest branch frequency samples. At roughly constant R_G, branched chains are higher in molecular weight than linear chains. Thus, retention in MTF increased with increasing levels of branching. The fractionation provided by MTF combined with the structural information attainable with on line multiangle light scattering detection will potentially enable determination of a long chain branching distribution for each hydrodynamic size slice or fraction. Coupling of MTF and SEC in an automated on line fashion would facilitate determination of long chain branching distributions across the molecular weight distribution of a sample. Present state-of-the-art technologies such as SEC with molecular weight sensitive detectors or 13C NMR provide only an average number of branches per chain. Thus, the determination of branching distributions across the molecular weight distribution would provide structural information that is not attainable at present.

In this paper, new columns for MTF were introduced. The columns were packed with submicron polydisperse silica that had been surface treated with PS. MTF elution order reversal was observed for PS standards and PE fractions injected onto these new columns. Additionally, M_C was found to be dependent on packing particle size. Thus, MTF elution order reversal was not unique to the previously studied monolithic columns,[1] but rather, appears to be a general phenomenon attainable when the interstitial channels (i. e. macropores) in a column are of the appropriate dimensions, and when the flow rate is reduced.

Although the mechanism of MTF is not completely understood, two potential mechanisms are pinning (or entanglement) and polymer restriction followed by relaxation. Both may be operative in a MTF experiment. Both are expected to be sensitive to long chain branching topology.

[1] D. M. Meunier, P. B. Smith, S. A. Baker, *Macromolecules* **2005**, 38, 5313–5320.
[2] P. B. Smith, D. M. Meunier, S. A. Baker, R. K. Prud'homme, International Patent Application, Publication Number WO 03/098208 A1, November 27, **2003**.

[3] *NMR and Macromolecules*, J. C. Jr, Ed., American Chemical Society, Washington, D.C. **1984**, Chapter 9.
[4] Wood-Adams, P. M. Randall, J. M. Dealy, A. W. deGroot, O. D. Redwine, Macromolecules. **2000**, *33*, 7489–7499.
[5] C. W. Macosko, *Rheology, Principles, Applications and Measurements*, Wiley-VCH New York **1994**, page 505.
[6] K. Terao, J. W. Mays, *Eur. Polym. J.* **2004**, *40*, 1623–1627.
[7] H. Benoit, *J. Polym. Sci.* **1953**, *11*, 507–510.
[8] T. Altares, Jr., D. P. Wyman, V. R. Allen, K. Meyersen, *J. Polym. Sci: Part A* **1965**, *3*, 4131–4151.
[9] Personal communication with Rob Edam, University of Amsterdam, June, 2006.
[10] E. Grushka, *Anal. Chem.* **1972**, *44*, 1733.
[11] P. Tackx, J. C. J. F. Tackx, *Polymer* **1998**, *39*, 3109–3113.
[12] P. J. Flory, T. G. Fox, *J. Am. Chem. Soc.* **1950**, *73*, 1904.
[13] B. H. Zimm, W. H. Stockmayer, *J. Chem. Phys.* **1949**, *17*, 1301–1314.

Crystallization Elution Fractionation. A New Separation Process for Polyolefin Resins

B. Monrabal, J. Sancho-Tello, N. Mayo, L. Romero*

Summary: New crystallization procedures have been developed for the analysis of the chemical composition distribution in polyolefins by pumping a small flow of solvent during the crystallization cycle. One of the new techniques, crystallization elution fractionation (CEF) combines the separation power of TREF and CRYSTAF and has been shown to provide very fast analysis of the composition distribution.

Keywords: CCD; CRYSTAF; crystallization; polyethylene; TREF

Introduction

The introduction of single site catalysts and multiple reactor technology allows the design of new polyolefin resins with improved properties for each application, and the chemical composition distribution (CCD) is the most discriminating microstructure parameter in these polymers. The proper analysis of the CCD necessarily requires a fractionation process, and as the incorporation of comonomer will result in the presence of branches or functional groups, and those will influence the crystallinity, it seems obvious that existing techniques like temperature rising elution fractionation and crystallization analysis fractionation were based on a fractionation step according to crystallizability.

Temperature Rising Elution Fractionation

Temperature rising elution fractionation (TREF) has been the most used analytical approach for the analysis of the CCD since its introduction in the industrial practice by Wild et al.[1] to characterize linear low density polyethylene. TREF analysis resembles a liquid chromatography separation where the sample is first dissolved in a proper solvent at high temperature and the solution is then introduced into a column containing an inert support; this is followed by a crystallization step at a slow cooling rate with no flow, during which polymer fractionation occurs by segregation of crystal aggregates of decreasing crystallinity as temperature goes down. Fractionation takes place within this cycle without physical separation of the fractions; all the crystal aggregates, from different crystallinity or branch content are still mixed together and are being deposited *in situ* on the same spot of the column where the initial polymer solution has been loaded.

TREF still requires a second temperature cycle to physically separate or quantify those fractions. This is achieved by pumping new solvent while the temperature is being increased. The eluant dissolves fractions of increasing crystallinity, or decreasing branch content, as temperature rises. These fractions are collected (preparative TREF) or their concentration monitored with an infrared detector (analytical TREF) to generate the CCD curve. The name temperature rising elution fractionation derives from this second temperature cycle.

Although analytical conditions are performed far from thermodynamic equilibrium, it has been shown that the elution temperature still follows a linear relation with the molar fraction of comonomer incorporated (m_c) as predicted from Flory equation for copolymers,[2] that after proper simplification can be reduced to

Polymer Characterization, c/Nicolás Copérnico 10, Valencia Technology Park, E-46980 Paterna, Spain
E-mail: bmonrabal@polymerchar.com

Equation (1):

$$T_m \cong T_m^0 - \frac{R(T_m^0)^2}{\Delta H_u} \cdot mc \qquad (1)$$

where T_m is the equilibrium melting temperature of the polymer-diluent mixture, T_m^0 the melting temperature of the pure polymer, and ΔH_u the heat of fusion per polymer repeating unit. Equation (1) assumes that ΔH_u is constant in the crystallization temperature range and that the presence of solvent, when crystallizing in solution, plays as an additional temperature shift factor.

Analysis of the CCD by TREF is today a common practice in the polyolefin industry and the long analysis time of the first homemade instruments[1,3] (three or four days per sample) has been reduced significantly down to a few hours; still, there is an interest in further reducing TREF analysis time. Reviews of the TREF technique have been done by Wild,[4] Glöckner,[5] Fonseca and Harrison,[6] Soares and Hamielec,[7] Anantawaraskul, Soares and Wodd-Adams,[8] and Monrabal.[9]

Crystallization Analysis Fractionation

Crystallization analysis fractionation (CRYSTAF) was developed by Monrabal[10,11] in 1991 as a process to speed up the analysis of the CCD, and it shares with TREF the same fundamentals on separation according to crystallizability, but the whole fractionation process is carried out during crystallization.

In CRYSTAF the analysis is carried out in stirred crystallization vessels with no support, by monitoring the polymer solution concentration, through the crystallization process, while decreasing temperature. Aliquots of the solution are filtered (through an internal filter inside the vessel) and analyzed by a concentration detector. In fact, the whole process is similar to a classical stepwise fractionation by precipitation with the exception that, in this approach, no attention is paid to the precipitate but to the polymer that remains in solution.

The first data points, taken at temperatures above any crystallization, provide a constant concentration equal to the initial polymer solution concentration; as temperature goes down the most crystalline fractions, composed of molecules with zero or very few branches (highly crystalline) will precipitate first, resulting in a steep decrease in the solution concentration on the cumulative plot. This is followed by precipitation of fractions of increasing branch content (or less crystallinity) as temperature continues to decrease; the last data point, corresponding to the lowest temperature of the crystallization cycle, represents the fraction which has not crystallized (mainly highly branched or amorphous material) and remains in solution at the lowest temperature. The first derivative of this curve corresponds to the CCD, very similar in shape to the one obtained in TREF with the only difference of a temperature shift, as equilibrium is not reached, and CRYSTAF is measured in the crystallization while TREF is measured in the dissolution (melting).

With this approach, the CCD can be analyzed relatively fast in a single crystallization cycle without physical separation of the fractions. The term crystallization analysis fractionation stands for this process. Reviews of the CRYSTAF technique have been done by Soares and Hamielec,[7] Anantawaraskul et al.[8] and Monrabal.[9]

Dynamic Crystallization

In the previous sections, TREF and CRYSTAF methods have been reviewed, and it has been discussed how both techniques share the same principles of fractionation on the basis of crystallizability through a slow cooling of a polymer solution. TREF is carried out in a packed column and demands two full temperature cycles, crystallization and elution, to achieve the analysis of the composition distribution. In CRYSTAF the analysis is performed in a single step, the crystallization cycle.

In Figure 1.a, the analysis of a blend of three different components by TREF is represented in three steps: 1) Sample loading into the column, 2) Crystallization cycle where the components are being crystallized in the same location where

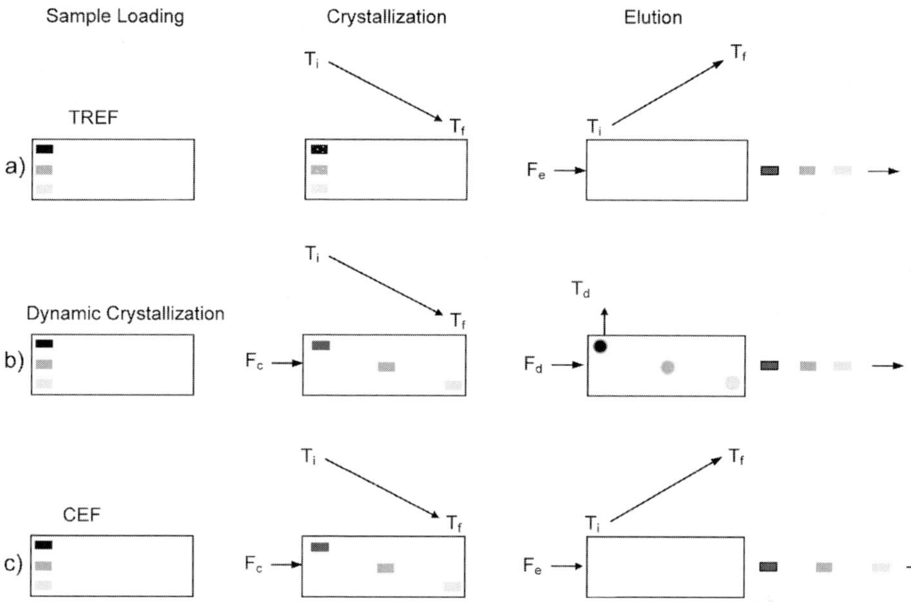

Figure 1.
Separation diagram by crystallizability. a) TREF separation process, b) Dynamic crystallization, c) Crystallization elution fractionation.

sample was loaded, and 3) Elution cycle where a solvent flow, F_e, is passed through the column to elute the fractions being dissolved as the temperature rises. The three components of the blend are physically separated from each other in the last cycle as shown in Figure 1.a.

A new separation approach is presented here based on the same principles of crystallizability and using a packed column like in TREF, but performing the physical segregation of fractions in the crystallization step as in CRYSTAF. The new separation process is known as dynamic crystallization because the cooling is performed while a small flow of solvent is passed through the column.

Dynamic crystallization is also a three-step procedure as represented in Figure 1.b, but physical separation of the components takes place within the column itself in the crystallization cycle. In dynamic crystallization, a small solvent flow, F_c, is passed through the column during crystallization, in such a way that when a component reaches its crystallization temperature it is segregated and anchored on the support while the other components, still in solution, move along the column until they reach their own crystallization temperature. At the end of the crystallization cycle, the three components are separated inside the column according to crystallizability.

The flow rate in during crystallization, F_c, plays an important role as it has to be adapted to the crystallization rate, C_r, crystallization temperature range, ΔT_c, of the components to be separated, and column volume, V_c, in order to crystallize all the components within the column length; F_c is calculated as follows from Equation (2):

$$F_c = \frac{V_c}{\Delta T_c} \times C_r \qquad (2)$$

Once the crystallization cycle is completed, the flow is interrupted and the column is heated for a few minutes at a temperature where all components are dissolved. After that, the elution cycle begins by adding a proper elution flow to the column. The separation order of the components will be according to CRYSTAF rather than TREF elution temperatures, as separation occurred in the crystallization and not in the dissolution process.

Proper analysis by this technique demands that the temperature range of the polymer components to be separated, ΔT_p is within the crystallization temperature range ΔT_c, and the full power of the dynamic crystallization analysis will be obtained when ΔT_p is equal to the ΔT_c used, resulting in the components crystallizing along the whole length of the column. When the range of ΔT_p is narrower and within the range of ΔT_c, only a fraction of the separation power, equivalent to the ratio of the two temperature ranges, will be obtained.

Crystallization Elution Fractionation

Once dynamic crystallization was developed as a new separation process, it was easy to realize the possibility to combine this crystallization step with a final elution cycle as in TREF to obtain a new *extended* separation. This is represented in Figure 1.c where the dynamic crystallization cycle, at a crystallization solvent flow of F_c, is followed by the temperature rising elution cycle, with a solvent flow F_e as in TREF. The new process is known as crystallization elution fractionation (CEF), as it combines the separation obtained in the crystallization step with the one obtained in the elution cycle; this is schematically represented in Figure 1.c by the extended separation of the three components at the exit of the column in CEF analysis as compared to the TREF approach.

It is quite interesting that the separation power of CRYSTAF (obtained by dynamic crystallization) and TREF are combined in CEF when both systems are based on the same crystallizability principles, and when TREF requires both cycles *per se*. On the other hand, one could expect that analysis of particular blends where differences in supercooling result in separation of the components during crystallization in an opposite direction to the one obtained by elution-melting would result in poor CEF separation.

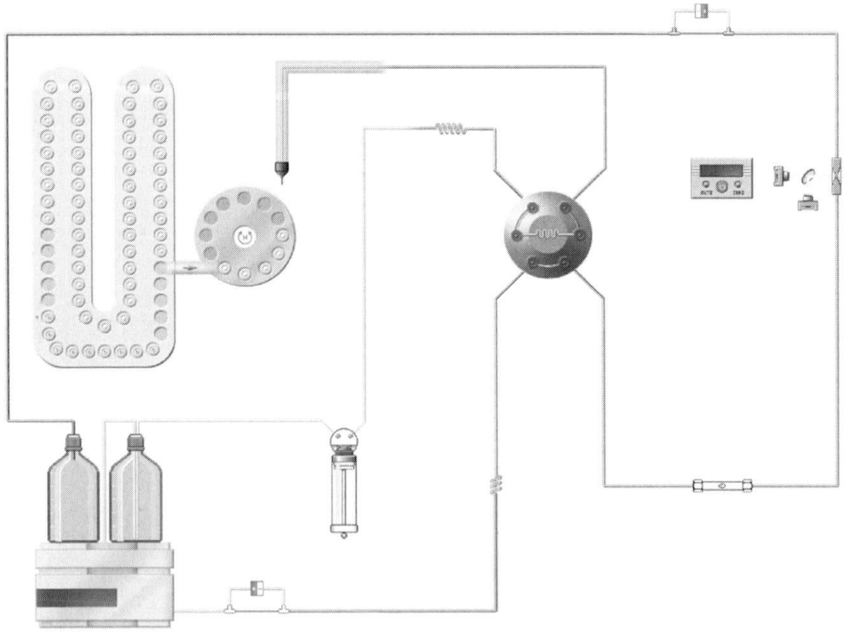

Figure 2.
Schematic diagram of the combined CEF, dynamic crystallization and TREF instrument.

Experimental Part

Dynamic crystallization and CEF experiments were performed in a specially developed apparatus constructed around a simplified Polymer ChAR TREF instrument. As the interest of CEF technology in these initial stages was focused on high throughput analysis, the instrument built incorporated an autosampler for 70 vials of 10 ml.

The CEF instrument diagram is quite simple as shown in Figure 2. The autosampler dissolves the sample in o-dichlorobenzene and it is loaded into the injection valve loop through the syringe dispenser. The sample is injected with the pump flow into the column head and the dynamic crystallization process begins at a given cooling rate and crystallization flow. As the crystallization ends, the oven starts the heating program and flow is adapted to the elution flow (usually higher than crystallization flow) passing through the column to a dual wavelength infrared detector, so concentration and composition can be measured at once. A dual capillary viscometer, as shown in the diagram, was added to the system to measure the composition – molar mass dependence.

The same instrument can be programmed to run TREF, dynamic crystallization, or CEF analysis as described squematically in Figure 1.

Results

Dynamic Crystallization

A blend of two metallocene resins of densities 0.902 and 0.937 g/ml was used to test the separation power of dynamic

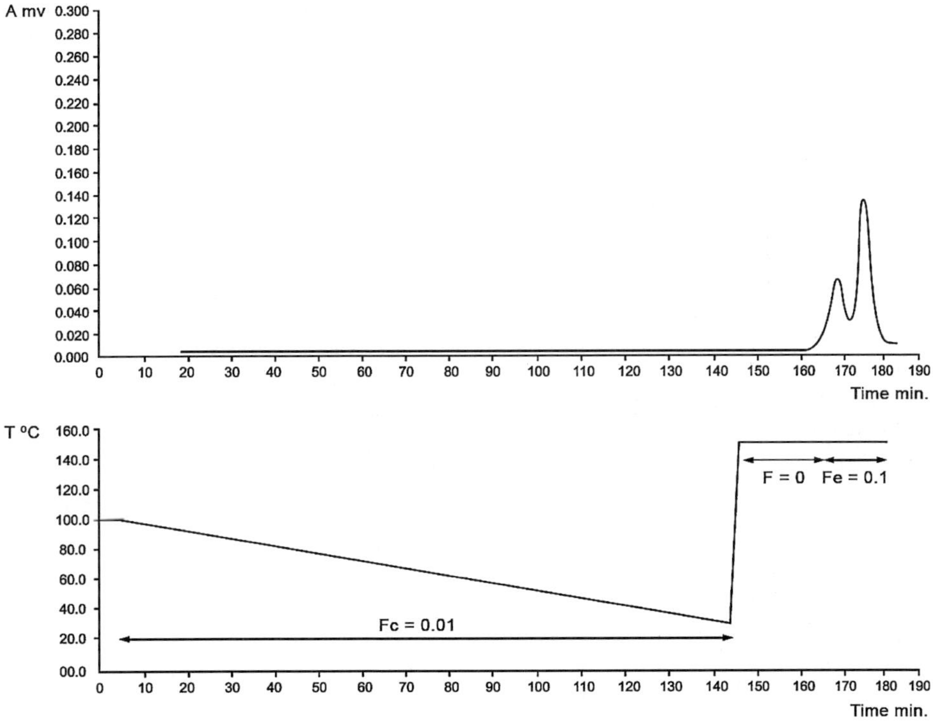

Figure 3.
Dynamic crystallization analysis of a blend of two metallocene resins. The top diagram is the elution curve showing the separation of the two components. The bottom diagram is the temperature profile with the flow being used at each step.

crystallization analysis. The sample at a concentration of 0.8% w/v was dissolved in a 10 ml vial at the autosampler and 0.1 ml solution was injected into the column. The crystallization rate was 0.5 °C/min. The crystallization flow calculated by Equation (2), according to the column interstitial volume, was 0.01 ml/min. Once the crystallization ends, the flow is stopped and the oven goes in a fast ramp up to 150 °C, staying for ten minutes to ensure full dissolution of the components previously crystallized in the column. After that time, the elution flow begins at 0.1 ml/min to elute the two metallocene resins separated into the column as shown in Figure 3. When the same experiment was repeated without any crystallization flow, only one peak is obtained.

Crystallization Elution Fractionation

Most work done with crystallization elution fractionation so far has been directed to reduce analysis time for applications in high throughput screening of catalyst formulations. The CEF analysis of a complex polyolefin, Elite™ (trademark of the Dow Chemical Company) is shown in Figure 4 with a total analysis time of 23 minutes. The analysis was performed at very fast crystallization and heating rates (10 °C/min), and still the separation of the three components of this resin is reasonably good as shown by the infrared transmission plot in Figure 4 and the final calculated composition distribution in Figure 5.

The capabilities of the CEF technique for high throughput screening were tested analyzing automatically the same sample in 10 different vial preparations every half an hour at a crystallization rate of 5 °C/min and heating rate of 10 °C/min. The samples were previously dissolved at 150 °C for 30 minutes with gentle shaking in the autosampler. The results presented in Figure 6 show the good repeatability of the CEF analysis.

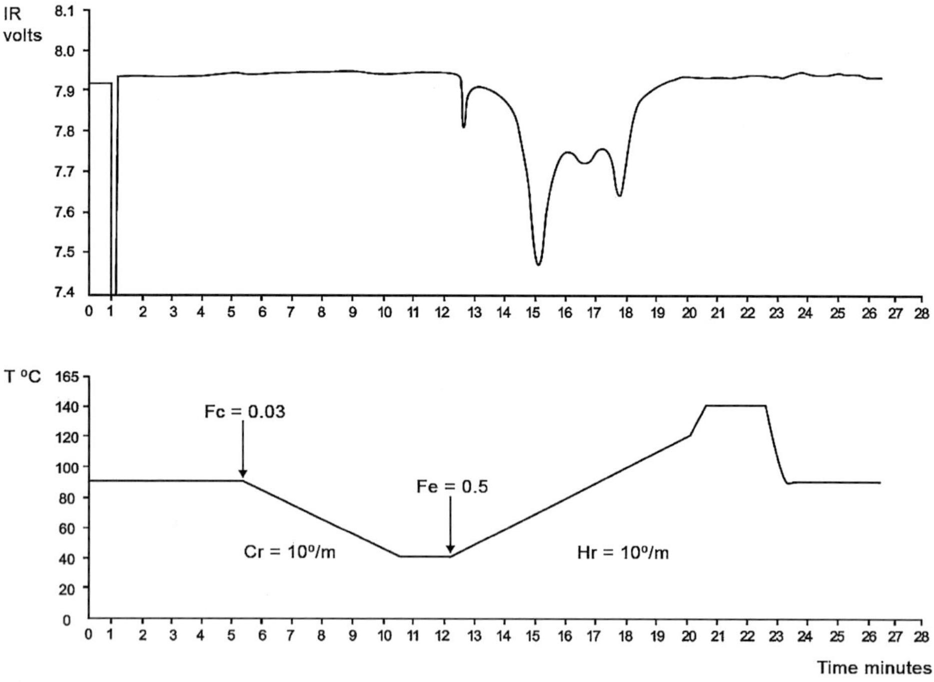

Figure 4.
CEF analysis of an Elite™ resin performed at fast crystallization and heating rates of 10 °C/min. The top diagram shows the elution infrared transmission curve. The bottom diagram is the crystallization and elution temperature profile.

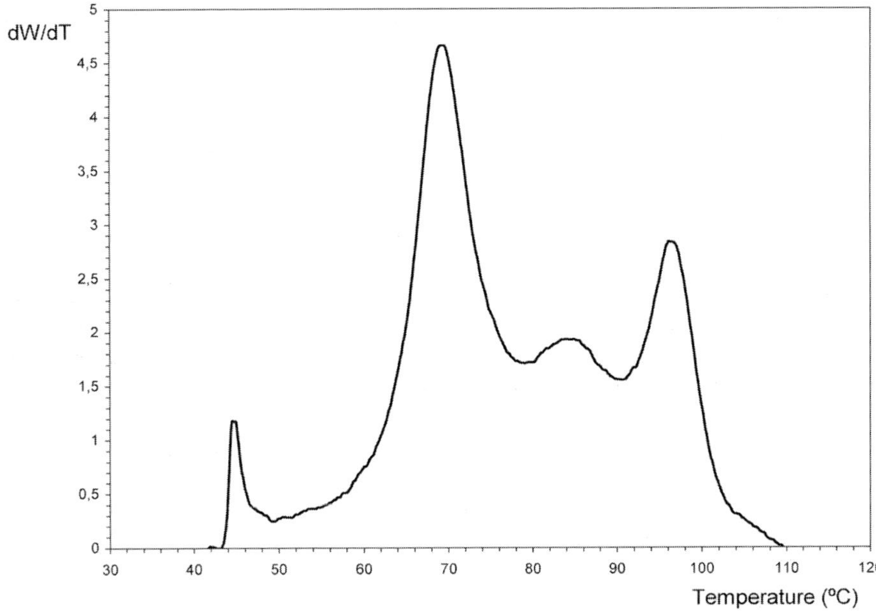

Figure 5.
CEF analysis of an Elite^{TM} resin obtained at crystallization and heating rates of 10 °C/min, injection volume of 20 μl (0.5%w/v) and elution flow rate of 0.5 ml/min.

Comparison of CEF and TREF

A blend of two metallocene resins of densities 0.902 and 0.916 g/ml was used to compare the separation of CEF and TREF under fast analytical conditions with the same column. In both cases the crystallization and heating rates were 2 °C/min and the elution flow rate 0.2 ml/min. The

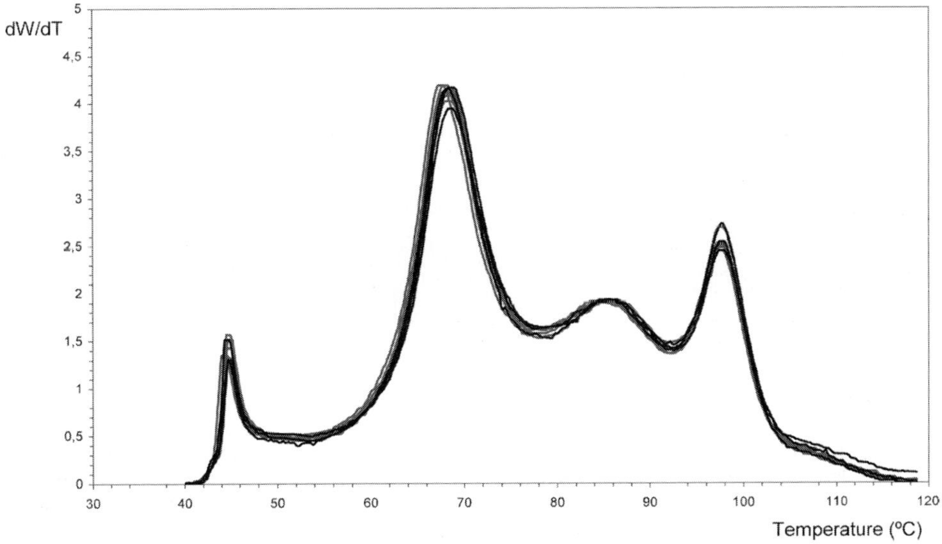

Figure 6.
Multiple CEF analysis (×10) of an Elite^{TM} resin obtained at crystallization rate 5 °C/min and heating rate 10 °C/min, injection volume of 20 μl (0.5%w/v), and elution flow rate of 0.5 ml/min.

only difference between CEF and TREF analysis was the crystallization flow of 0.12 ml/min present in the CEF analysis. The results presented in Figure 7 show the improved resolution of CEF over TREF analysis in the separation of the two components.

Multiple Detectors

The incorporation of other detectors in line to CEF analysis is as easy as with GPC or TREF techniques. In the initial experiments, a dual capillary viscometer and a composition sensor were installed in the CEF instrument to evaluate their potential. The analysis of a linear low density polyethylene is shown in Figure 8. The ratio of CH_3 over CH_2 signals (B/A plot) results in the gray composition line (methyls per 1000 carbon atoms) increasing linearly, as expected, towards lower temperatures. The ratio of viscometer over concentration signals (C/A) results in the intrinsic viscosity black line that increases towards higher temperatures (lower branch content) as expected in a Ziegler-type resin.

Conclusions

A new dynamic crystallization approach has been developed to separate semicrystalline polymers into a packed column according to crystallizability. The physical separation takes place in the crystallization cycle and therefore the results will correlate to CRYSTAF temperature data. Future work is required to better evaluate the interest of this new approach in industrial processes for analytical or preparative fractionation purposes.

Having developed dynamic crystallization, it was obvious to consider this approach to replace the static crystallization cycle being used in TREF before the elution step. The combination of dynamic crystallization and temperature rising elution crystallization results in a new techni-

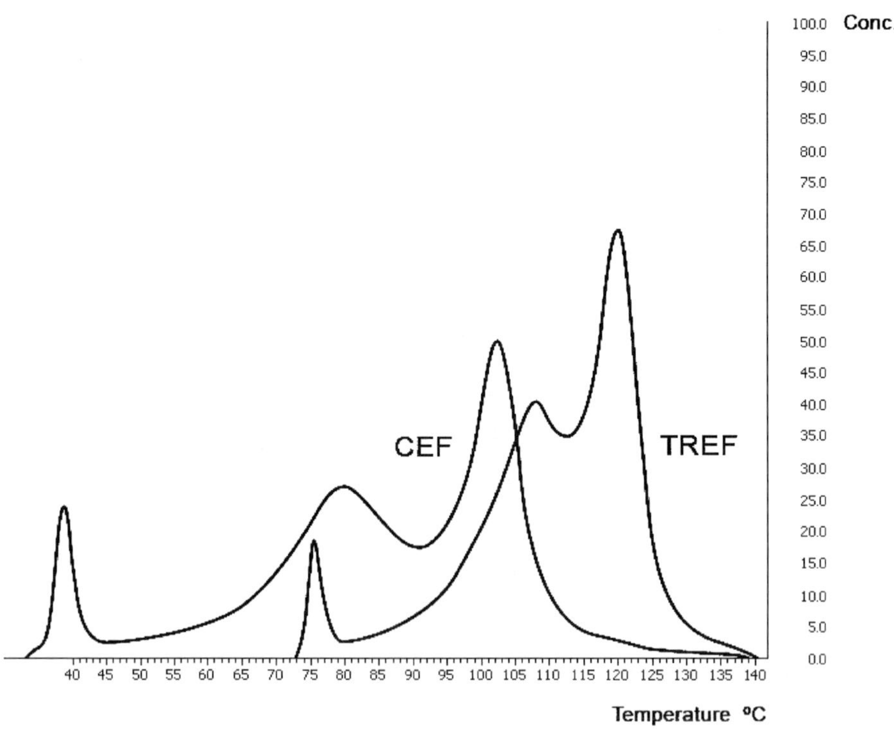

Figure 7.
Analysis of a blend of two metallocene resins by CEF and TREF.

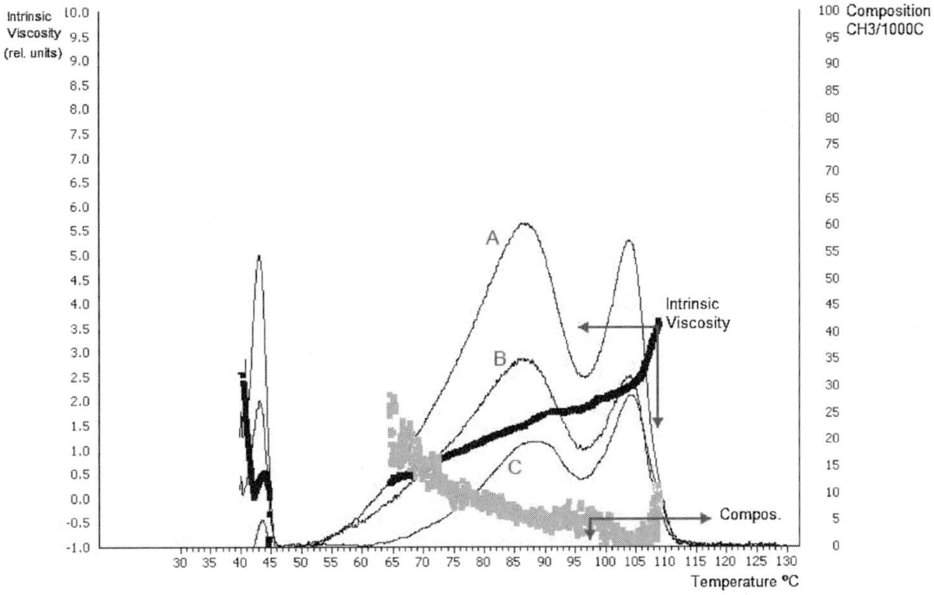

Figure 8.
CEF analysis of a linear low density polyethylene. A - Concentration (CH_2), B - Composition (CH_3) and C - Viscometer output.

que with extended separation. The new technique has been named crystallization elution fractionation, CEF, as it combines the separation power obtained in the crystallization cycle (equivalent to CRYSTAF), with the one obtained in the elution-melting cycle (equivalent to TREF).

CEF is performed in a simplified TREF instrument and can easily adapt viscometry, light scattering, composition, or other detectors to determine molar-mass composition dependence or to obtain further information of polymer microstructure. The same CEF apparatus can perform TREF and dynamic crystallization.

The analysis time of the CCD has been dramatically reduced with improvements in TREF column and hardware design, and with newer techniques becoming available. The analysis by classical TREF of one sample took 100 hours in the eighties (around five samples could be analyzed per week). The development of CRYSTAF in the nineties allowed analyzing five samples in 8 hours (15 samples per day). In the last years, improvements in TREF allowed the analysis of five to ten samples per day. Today, CEF can analyze one sample in less than 30 minutes, enough for the demanded high throughput screening of 50 samples per day.

[1] L. Wild, T. Ryle, *Polym. Prepr., Am. Chem. Soc., Polym. Chem. Div.* **1977**, *18*, 182.
[2] P. J. Flory, *Principles of Polymer Chemistry*, Cornell Univ. Press, Ithaca, N.Y. **1953**.
[3] L. Wild, T. Ryle, D. Knobeloch, *Polym. Prepr., Am. Chem. Soc. Polym. Chem. Div.* **1982**, *23*, 133.
[4] L. Wild, *Adv. Polym. Sci.* **1991**, *98*, 1–147.
[5] G. Glöckner, *J. Appl. Polym. Sci., Appl. Polym. Symp.* **1990**, *45*, 1–24.
[6] C. A. Fonseca, I. R. Harrison, *Modern techniques for Polymer Characterisation*, R. A. Pethrick, Ed., **1999**, p. 1–13.
[7] J. B. P. Soares, A. E. Hamielec, *Modern techniques for Polymer Characterisation*, R. A. Pethrick, Ed., **1999**, p. 15–55.
[8] S. Anantawaraskul, J. B. P. Soares, P. M. Wood-Adams, *Adv. Polym. Sci.* **2005**, *182*, 1–54.
[9] B. Monrabal, *Encyclopedia of Analytical Chemistry*, John Wiley & Sons, **2000**, p. 8074–8094.
[10] B. Monrabal, Crystallization Analysis fractionation, US Patent 5,222,390, **1991**.
[11] B. Monrabal, *J. Appl. Polym. Sci.* **1994**, *52*, 491.

 DOI: 10.1002/masy.200751107

Block Index for Characterizing Olefin Block Copolymers

Colin Li Pi Shan, Lonnie G. Hazlitt*

Summary: Olefin block copolymers produced by chain shuttling catalysis exhibit crystallinity characteristics that are distinct from what would be expected for typical random olefin copolymers with comparable monomer compositions produced from either 'single-site' or heterogeneous catalysis. Olefin block copolymers produced by chain shuttling catalysis have a statistical multiblock architecture. A unique structural feature of olefin-based block copolymers is that the intra-chain distribution of comonomer is segmented (statistically non-random). Fractionating an olefin block copolymer by preparative temperature rising elution fractionation, TREF, results in fractions that have much higher comonomer content than comparable fractions of a random copolymer collected at an equivalent TREF elution temperature. We have developed a "block index" methodology which quantifies the deviation from the expected monomer composition versus the analytical temperature rising elution fractionation, ATREF, elution temperature. When interpreted properly, this index indicates the degree to which the intra-chain comonomer distribution is segmented or blocked. The unique crystallization behavior of block copolymers determine the magnitude of the block index values because the highly crystalline segments along an otherwise non-crystalline chain tend to dominate the ATREF (and DSC) temperature distributions.

Keywords: ATREF; DSC; ethylene-alpha olefin copolymer; olefin block copolymer; random copolymer; TREF

Introduction

The recent break-through in chain shuttling catalyst technology has enabled the production of novel linear olefin multi-block copolymers via coordination catalysis in Dow's continuous solution process. The chain shuttling technology is based on two catalysts with varying comonomer incorporation capabilities and a shuttling agent which transfers a growing polymer chain from a good comonomer incorporator catalyst site to a poor comonomer incorporator catalyst site in a reversible manner. The good comonomer incorporator makes the soft (or non-crystallizing) segments and the poor comonomer incorporator makes the hard (or crystallizing) segments of the block copolymer. The polymerization and properties of these olefin block copolymers (OBCs) have been reviewed recently.[1,2]

This new type of olefin block copolymer is constructed from hard block segments that are rich in ethylene and soft block segments that have α-olefin comonomer along the same polymer chain. These block copolymers have relatively narrow MWD (M_w/M_n ~2) and contain multiple block segments that arise from statistical addition. The number and length of block segments are controlled by the concentration of chain shuttling agent present during the polymerization. In addition to controlling the overall density and melt index of the copolymers, the amount and composition of the segments can be controlled

Specialty Plastics and Elastomers R&D, The Dow Chemical Company, Freeport, TX
E-mail: cllipishan@dow.com

to tailor-make materials for specific applications.[3]

Figure 1 provides an example of the hypothetical microstructure of an OBC produced by chain shuttling catalysis. This type of block copolymer can be termed a "linear statistical multiblock" or LSMB and is quite different than the conventional block copolymers made by living polymerization chemistry as well as commercially available polyolefins. In a LSMB the chain lengths, comonomer distribution, and block distribution of block sizes and lengths are polydisperse. By contrast, the traditional block copolymers have structures that are well-controlled to be nearly mono-disperse (equal chain lengths) and the lengths of each type of block are essentially equal from chain to chain as dictated by the polymerization chemistry and conditions.

For typical random polyolefins, the molecular structure distributions such as the molecular weight distribution and/or the short chain branching distribution can be measured by GPC and ATREF/CRYSTAF, respectively. However, neither of these techniques is able to measure the blockiness of a copolymer (without prior knowledge of its structure), as reflected in the intrachain monomer distribution. To our knowledge, there are no analytical tools in the literature to specifically characterize an olefin block copolymer.

In this paper, we describe some of the unique analytical characteristics of an olefin block copolymer. In addition, based on deviations from the observed relationship for structure and ATREF elution temperature for random copolymers, we outline a "Block Index" method to quantify the observed microstructure and distinguish it from traditional random copolymers.

Experimental Part

Materials

An olefin block copolymer synthesized by the chain-shuttling method was produced in a continuous polymerization process as described by Arriola et al.[2] The olefin block copolymer was produced using ethylene and 1-octene comonomer as per the design shown in Table 1.

Material Analysis

Differential Scanning Calorimetry (DSC)

Differential scanning calorimetry was performed on a TA Instruments Q1000 DSC equipped with an RCS cooling accessory and an auto sampler. A nitrogen purge gas flow of 50 ml/min was used. The sample was pressed into a thin film and melted in the press at about 190 °C and then air-cooled to room temperature (25 °C). About 3–10 mg of material was then cut, accurately weighed, and placed in a light aluminum pan (ca. 50 mg) which was later crimped shut. The thermal behavior of the sample was investigated with the following temperature profile: the sample was rapidly heated to 190 °C and held isothermal for 3 minutes in order to remove any previous thermal history. For these octene-based polymers, the sample was then cooled to −40 °C at 10 °C/min cooling rate and held at −40 °C for 3 minutes. The sample was then heated to 150 °C at 10 °C/min heating rate. The cooling and second heating curves were recorded.

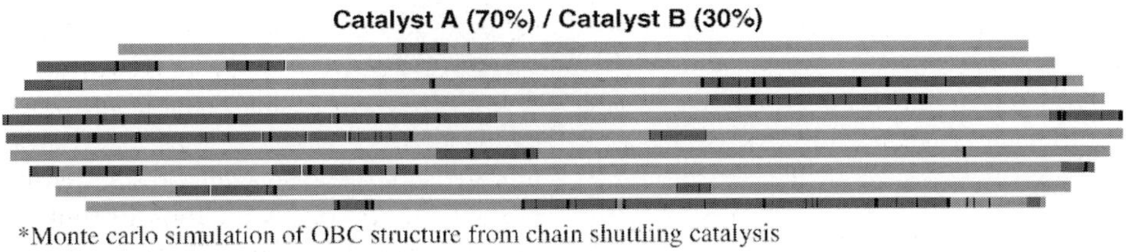

Figure 1.
A pictorial example of a linear statistical multiblock copolymer.

Table 1.
Design of a olefin block copolymer.

Example	Density (g/cc)	I2 (g/min)	wt% Hard Segment or polymer in blend	Targeted* wt% Octene Hard Segment or polymer in blend	Targeted* wt% Soft Segment or polymer in blend	Targeted* wt % Octene in Soft Segment or polymer in blend	Chain Shuttling Agent Present
OBC	0.8786	1.5	29	0.4	71	15	Yes
In Reactor Blend	0.8895	0.9	39	0.4	61	15	No

* As controlled by the polymerization conditions via predictive reactor modelling.

Analytical Temperature Rising Elution Fractionation (ATREF)

In ATREF analysis, the composition to be analyzed was dissolved in 1,2,4 trichlorobenzene (2 mg/ml) at 160 °C and allowed to crystallize in a column containing an inert support (stainless steel shots) by slowly reducing the temperature to 20°C, at a rate of −0.1 °C/min, where the temperature was held for one hour. The instrumentation was equipped with an IR4 infra-red detector (PolymerChar, Valencia, Spain). An ATREF chromatogram curve was then generated by eluting the crystallized polymer sample (at a flow rate of 1 cc/min) from the column while slowly increasing the temperature of the column and eluting solvent (trichlorobenzene) from 20 to 120°C (at a rate of 1.0°C/min).

^{13}C Nuclear Magnetic Resonance (NMR) Analysis

The samples were prepared by adding approximately 3 g of a 50/50 mixture of tetrachloroethane-d^2/orthodichlorobenzene to 0.4 g sample in a 10 mm NMR tube. The samples were then dissolved and homogenized by heating the tube and its contents to 150 °C. The data was collected using a JEOL Eclipse™ 400 MHz spectrometer or a Varian Unity Plus™ 400 MHz spectrometer, corresponding to a ^{13}C resonance frequency of 100.5 MHz. The data was acquired using 4000 transients per data file with a 6 second pulse repetition delay. To achieve maximum signal-to-noise for quantitative analysis, multiple data files were added together. The spectral width was 25,000 Hz with a minimum file size of 32K data points. The samples were analyzed at 130 °C in a 10 mm broad band probe. The comonomer incorporation was determined using the method reported by Zhou et al.[4]

Polymer Fractionation by TREF

Large-scale TREF fractionation was carried out by dissolving 15–20 g of polymer in 2 liters of 1,2,4-trichlorobenzene (TCB) by stirring for 4 hours at 160 °C. The polymer solution is forced by 15 psig (100 kPa) nitrogen onto a 3 inch by 4 foot (7.6 cm × 121.9 cm) steel column packed with a 60:40 (v:v) mix of 30–40 mesh (600–425 μm) spherical, technical quality glass beads (available from Potters Industries, HC 30 Box 20, Brownwood, TX, 76801) and stainless steel, 0.028″ (0.7mm) diameter cut wire shot (available from Pellets, Inc. 63 Industrial Drive, North Tonawanda, NY, 14120). The column is immersed in a thermally controlled oil jacket, set initially to 160 °C. The column is first cooled ballistically to 125 °C, then slow cooled to 20 °C at 0.04 °C per minute and held for one hour. Fresh TCB is introduced at about 65 ml/min while the temperature is increased at 0.167 °C per minute. Approximately 2000 ml portions of eluant from the preparative TREF column are collected in a 16-station, heated fraction collector. The polymer is concentrated in each fraction using a rotary evaporator until about 50 to 100 ml of the polymer solution remains. The concentrated solutions are allowed to stand overnight before adding excess methanol, filtering, and rinsing (approx. 300–500 ml of methanol including the final rinse). The filtration step is performed on a 3 position vacuum assisted filtering station using 5.0 μm polytetrafluoroethylene coated filter paper (available from Osmonics Inc., Cat# Z50WP04750). The filtered fractions are dried overnight in a vacuum oven at 60 °C and weighed on an analytical balance before further testing.

Results and Discussion

Figure 2 shows that olefin block copolymers produced by chain shuttling catalysis have a relatively constant melting-point-versus-density relationship. Karande et al.[3] have shown that OBCs exhibit elastomeric character while still retaining a high melting point (115–120 °C) which is attributed to the presence of HDPE-like hard segments. However, the observation of a high melting point alone is not sufficient to confirm a polymer's blocky nature. When compared to random copolymers, OBCs in the solid-state have a unique crystal morphology and show improved physical properties such as compression set, elastic recovery, and abrasion resistance.[3,5]

Figure 3 shows the analytical TREF profile comparing an olefin block copolymer (0.878 g/cc, 1.5 I_2) to a random copolymer (AFFINITY^{TM} VP8770, 0.887 g/cc, 0.9 I_2) and a polymer blend (0.89 g/cc, 1.0 I_2) with components that are representative of the hard and soft segments within the olefin block copolymer. Although not a perfect comparison, the differences shown in Figure 3 cannot be reconciled by the slight differences in density and melt index of these polymers. Table 2 summarizes the analytical characteristics of these polymers. Figure 3 shows that for this particular OBC, 90 wt% of the polymer eluted at a peak temperature of 80 °C. The nearly complete elution of this OBC, despite being 0.878 g/cc (19 wt% crystallinity) is unique when compared to the blend and random copolymer that have peak elution percentages of 35 wt% and 75 wt%, respectively. For the 0.889 g/cc blend, the 35 weight % of the ATREF peak is consistent with the targeted amount of high crystallinity polymer made. The fraction eluting below 30 °C

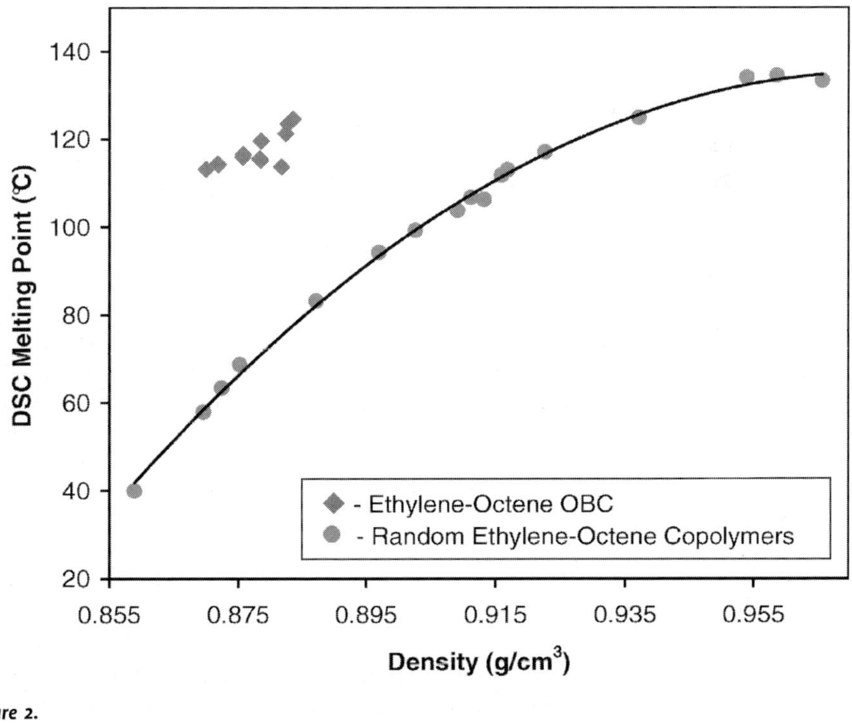

Figure 2.
Melting point characteristic of olefin block copolymers.

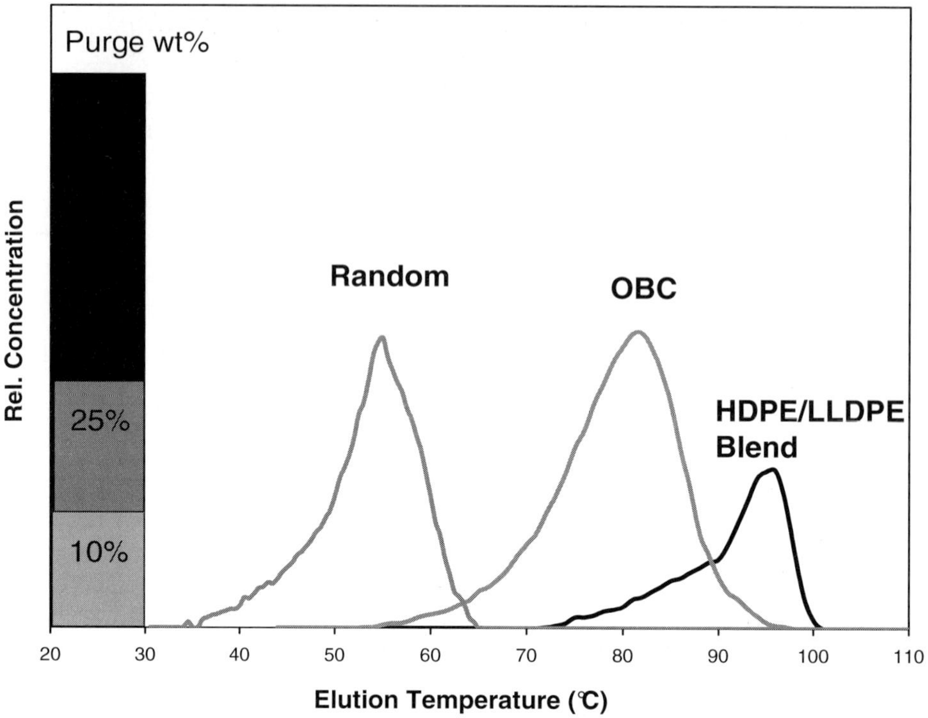

Figure 3.
Analytical TREF profile of an OBC, a random copolymer, and a polymer blend.

is referred to as the non-crystalline or "purge" fraction and contains the soft, low crystallinity polymer made. The 55 °C elution temperature and 75 weight percent peak area for the VP8770, random copolymer is consistent with its 0.887 g/cc density and 0.9 I_2. For the OBC, the area of the eluted peak was significantly higher than the targeted amount of hard segments in the overall polymer (29 wt% hard segments for this particular OBC). This indicates that the hard segments dominate the elution behavior of the OBC and are present in the backbone of the chains (supporting the contention that the soft, lower crystallinity segments are connected to the hard, higher crystallinity segments, which provides an indication of its blocky nature).

To confirm this observation, preparative TREF fractionation was carried out. Fractions were eluted in 5 °C increments and then analyzed for their octene content by NMR. Figure 4 and Tables 3–4 shows the expected behavior of the fractions prepared from an LLDPE produced by Ziegler-Natta catalysis (ATTANE™ 4203, 0.90 g/cm^3, 0.8 I_2) and an LLDPE produced by

Table 2.
Analytical characteristics of the polymers studied.

Example	Density (g/cc)	Mol % Octene	I2	I10/I2	M_w	M_w/M_n	Heat of Fusion	Crystallinity	T_m	T_c	ATREF Elution T	wt% Peak
	(g/cc)	(mol %)			(g/mol)		(J/g)	Wt%	(°C)	(°C)	(°C)	wt%
Random (VP8770)	0.8872	9.1	0.9	8.2	98000	2.2	74.8	25.6	83.2	65.1	55	75
Blend	0.8895	9.3	0.9	13.4	137300	13.8	90	30.8	125	111	82	35
OBC	0.8786	10.8	1.5	6.7	104600	2.0	55	18.8	120	101	96	90

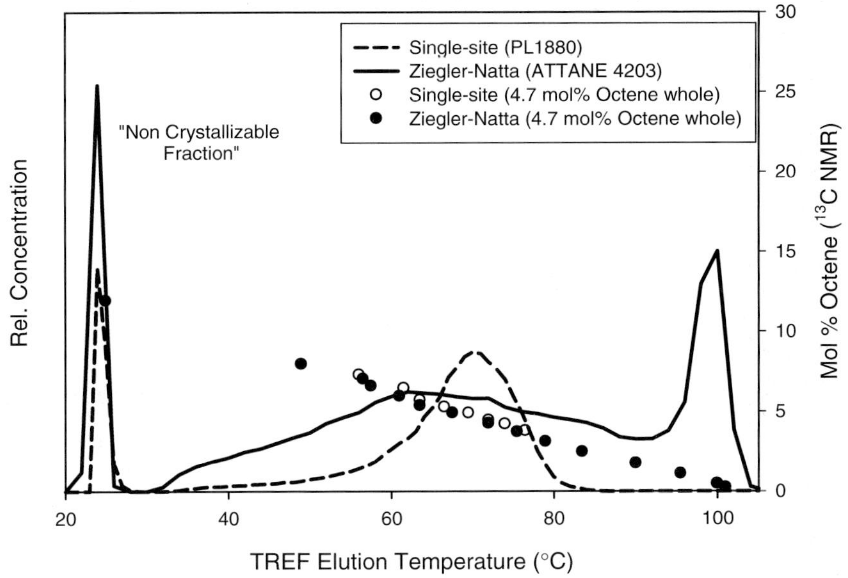

Figure 4.
Comonomer versus elution temperature relationship for random copolymers.

single-site catalysis (AFFINITY™ PL1880, 0.90 g/cm^3, 1 I_2). As shown for these polymers produced with the same comonomer type, a distinct relationship between the elution temperature and comonomer content is observed, indicating that regardless of the catalyst nature, the fractions from these random copolymers, for all practical purposes, have a statistically random distribution of comonomer. Wild has demonstrated that the peak elution temperature directly relates to the degree of short chain branching in a copolymer.[6] Brull et al. has demonstrated that the melting point or crystallization point depression of propylene-alpha olefin copolymers occurs linearly with increasing comonomer content and type regardless whether the analysis was performed in the melt or solution.[7]

Table 3.
Ziegler-Natta LLDPE TREF fractionation results.

Example - Ziegler-Natta Ethylene-Octene Copolymer (ATTANE™ 4203, 0.90 g/cc, 0.8 I2)			
Fractionation Temperature (°C)	Weight Fraction Recovered	ATREF T (°C)	Mol % Octene (NMR)
20	0.08	20	12.0
20–35	0.02	46.5	10.0
35–40	0.050	49	8.0
40–45	0.035	56.5	7.0
45–50	0.047	57.5	6.6
50–55	0.067	61	6.0
55–60	0.079	63.5	5.4
60–65	0.105	67.5	4.9
65–70	0.104	72	4.3
70–75	0.110	75.5	3.7
75–80	0.089	79	3.1
80–85	0.062	83.5	2.5
85–90	0.045	90	1.7
90–95	0.041	95.5	1.1
95–100	0.056	100	0.5
100–105	0.007	101	0.2

Table 4.
Single-site LLDPE TREF fractionation results.

Example - Single-site Ethylene-Octene Copolymer (AFFINITY™ PL1880, 0.90 g/cc, 1.0 I2)			
Fractionation Temperature (°C)	Weight Fraction Recovered	ATREF T (°C)	Mol % Octene (NMR)
40–45	0.025	56	7.3
45–50	0.041	61.5	6.5
50–55	0.073	63.5	5.7
55–60	0.124	66.5	5.3
60–65	0.180	69.5	4.9
65–70	0.228	72	4.4
70–75	0.241	74	4.2
75–80	0.050	76.5	3.8

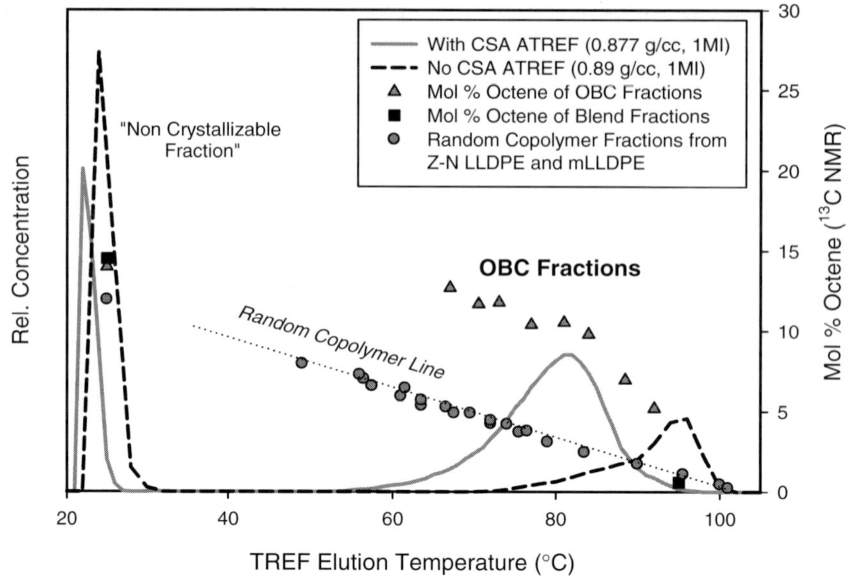

Figure 5.
Comonomer versus elution temperature of OBC.

Figure 5 and Tables 5–6 show the results for the OBC and in-reactor blend. Strikingly, it can be seen that for a given ATREF elution temperature, the octene content of the preparative fractions from the OBC copolymer is significantly higher than that of the preparative fractions from traditional random copolymers.

The expected comonomer composition versus elution temperature is shown by the calibration line for random copolymers. In terms of the distribution of comonomer in this OBC, it can be concluded that the comonomer distribution is statistically non-random and highly blocked (or segmented). Thus, for the high elution temperatures observed, the comonomer content of the fractions is significantly higher due to the presence of the octene-rich soft segment within the chain. In comparison, the octene content of the fraction from the in-reactor blend produced with no chain shuttling agent falls along the random copolymer line as expected.

Table 5.
OBC TREF fractionation results.

OBC TREF Fractionation (0.8786 g/cc, 1.5 I2)			
Fractionation Temperature (°C)	Weight Fraction Recovered	ATREF T (°C)	Mol % Octene (NMR)
20	0.162	20	14.1
20–60	0.104	67	13.7
60–65	0.040	70.5	12.6
65–70	0.053	73	12.2
70–75	0.126	77	11.7
75–80	0.220	81	10.5
80–85	0.192	84	9.8
85–90	0.065	88.5	7.0
90–95	0.019	92	5.2

Random Versus Block Copolymers

The observation that the TREF fractions from an OBC contain a higher than expected amount of comonomer demonstrates that the intrachain comonomer distribution or sequence distribution is

Table 6.
In-reactor blend TREF fractionation results.

BLEND TREF Fractionation (0.8895 g/cc, 0.9 I2)			
Fractionation Temperature (°C)	Weight Fraction recovered	ATREF T (°C)	Mol % Octene (NMR)
20	0.63	20	14.5
20–105	0.37	96	0.6

non-random and is in fact blocked, because it deviates from that of a statistically random copolymer in a theoretically predictable manner. Since the amount of comonomer can be predicted from a given elution temperature for a statistically random copolymer, the "blockiness" of an olefin block copolymer can be defined on the basis of Flory's equilibrium crystallization theory.[8] For random copolymers, Flory proposed a theoretical relationship for the probability of having a minimum crystallizable sequence length determined primarily by the equilibrium crystallization temperature. The conditional probability of sequential monomer insertion (ethylene in the case of the OBCs described here) is defined as *p*. The magnitude of the conditional probability, *p*, is compared to the molar fraction of monomer in the whole polymer, X_A, (refer to Figure 6) to test whether, based on a random probability of the intrachain monomer distribution, the copolymer is random. If the conditional probability of observing a crystallizable sequence is less than that determined by average molar monomer content of the whole polymer, then the polymer is described as having an alternating distribution. If the probability of observing a crystallizable sequence was greater than that determined by the average molar monomer content of the whole polymer, then the polymer is described as having a blocky

Flory's Probability Definitions

Sequence Propagation of A (in this case Ethylene):

p = is the probability that an A unit is succeeded by another A unit without being influenced by the nature of the preceding unit, in a sequence

X_A = mole fraction of crystallizable units (ethylene units)

The intrachain segment distribution can be shown to be fully determined by the mole fraction of each monomer for randomly incorporated chains. Assume X_A = 0.9 (90 mole % ethylene incorporation) and the following chains were selected at random from all available chains.

*previous ethylene unit inserted

$p = X_A$ (***Random***)

AAAAAAA**BB**AAAAAAAAAA**B**AAAAAAAAA**B**AAAAAAAAAAAA**B**AAAAAAA

$p < X_A$ (***Alternating***)

AAAAAAAAA**B**AAAAAAAAA**B**AAAAAAAAA**B**AAAAAAAAA**B**AAAAAAAAA**B**

$p > X_A$ (***Blocky***)

AAAAAAAAAAAAAAAAA**BBB**AAAAAAAAAAAAAAAAAA**BB**AAAAAAAAAA

Therefore, if you observe a difference between the conditional probability, p, and the mole fraction of ethylene, X_A, then the intrachain distribution of comonomer within the polymer is not random.

Figure 6.
Flory's polymer probability definitions.

structure. Finally, if the conditional probability and molar fraction are equal, the copolymer is random.

In the discussion below, we chose to express the mole fraction of ethylene as P, with the understanding that it is only equivalent to the conditional probability, p, described above when the copolymer is random. From this point forward, we will consider the measured mole fraction of ethylene (X) from NMR or other technique for a polymer as an estimate of the conditional probability.

Having now proposed a "blocky" structure for the OBC, we are in a position to validate the proposed structure on the basis of Flory's equilibrium melting theory as extended to the ATREF elution temperature. The key relationship of Flory's theory is:[8]

$$\frac{1}{T_m} - \frac{1}{T_m^0} = -\left(\frac{R}{\Delta H_u}\right) \ln P \qquad (1)$$

In Equation (1) the mole fraction of crystallizable monomers, P, is related to the melting temperature, T_m, of the copolymer, and the melting temperature of the pure crystallizable homopolymer, T_m^0. The equation is similar to the experimentally determined relationship for the natural logarithm of the mole fraction of ethylene as a function of the reciprocal of the ATREF elution temperature or melting point (°K) as shown in Figure 7 for various homogeneously branched copolymers of ethylene. *Note: The ATREF Elution Temperature-NMR calibration depicted in* Figure 7 *is not intended to be "universal" in that ATREF instrumentation and methodology varies considerably in the literature. The calibration in this paper is therefore to be considered to be unique and must be repeated for each ATREF instrument. Also, the preparative TREF fractions or homogeneous random copolymers that were chosen have weight average molecular weights of at least 100,000 daltons and polydispersity values of no greater than about 2.5.*

Preparative TREF fractions of nearly all random copolymers and their blends fall on this line, except for the small, predictable effects of molecular weight. According to Flory's definition, if one observes a difference between the measured (X) and predicted value (P) of the mole fraction of

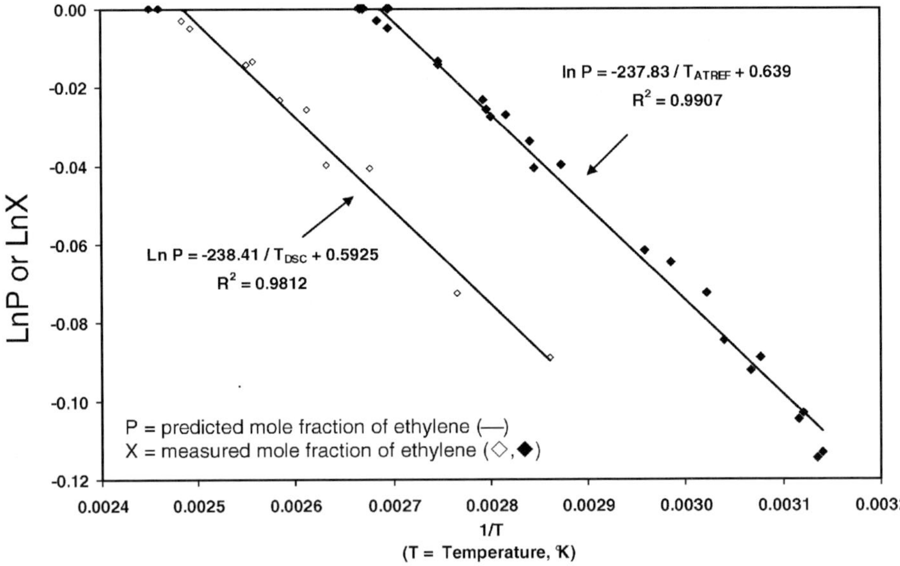

Figure 7.
The relationship of ethylene mole fraction to ATREF peak elution temperature (◆) and DSC melting temperature (◇) for various homogeneously branched copolymers can be interpreted on the basis of Flory's equation.

ethylene in an unknown polymer, then the intrachain distribution of comonomer within the polymer is not random.

A TREF fraction of a blocky copolymer should have more comonomer (octene) than a comparable random copolymer fraction with the same ATREF elution temperature (or DSC melting temperature). Figure 8 shows that TREF fractions from the OBC all lie below the random copolymer line when plotted as ln X versus $1/T_{ATREF}$. Thus, they have more octene than their random counterparts at the same TREF elution temperature.

Block Index

The mole fraction of ethylene in random copolymers is determined by the underlying statistic(s) of the ethylene segment distribution produced during the polymerization. The crystallization behavior, and ultimately the minimum equilibrium crystal thickness at a given crystallization temperature and melting temperature provides a useful way to at least qualitatively measure how "blocky" a given TREF fraction is relative to its random equivalent copolymer (or random equivalent TREF fraction). In Figure 8, the ln X versus $1/T_{ATREF}$ relationship shows the magnitude of the deviation of an OBC fraction from the expected random relationship. However, it is important to realize that there are two random equivalents for a given blocky fraction, one corresponding to constant elution or melting temperature and one corresponding to constant mole fraction of ethylene. These form the sides of a right triangle as shown in the Figure 9.

For any of the fractions in Figure 9, the distance to the random copolymer line is easily calculated using Pythagorean Theorem with the guidance of Figure 9.

In Figure 9, the T_X and X_X values are the ATREF elution temperature and the ethylene mole fraction measured by NMR, respectively, for an arbitrary preparative TREF fraction from an olefin block copolymer. The T_A and P_A values are the ATREF elution temperature and the ethylene mole fraction for the pure "hard segment". This point can be set to values for high density polyethylene homopolymer or it can be set to values corresponding to the actual hard segment, if known. The P_{AB} value corresponds to the measured (NMR) ethylene mole fraction in the whole polymer prior to fractionation and the T_{AB} value corresponds

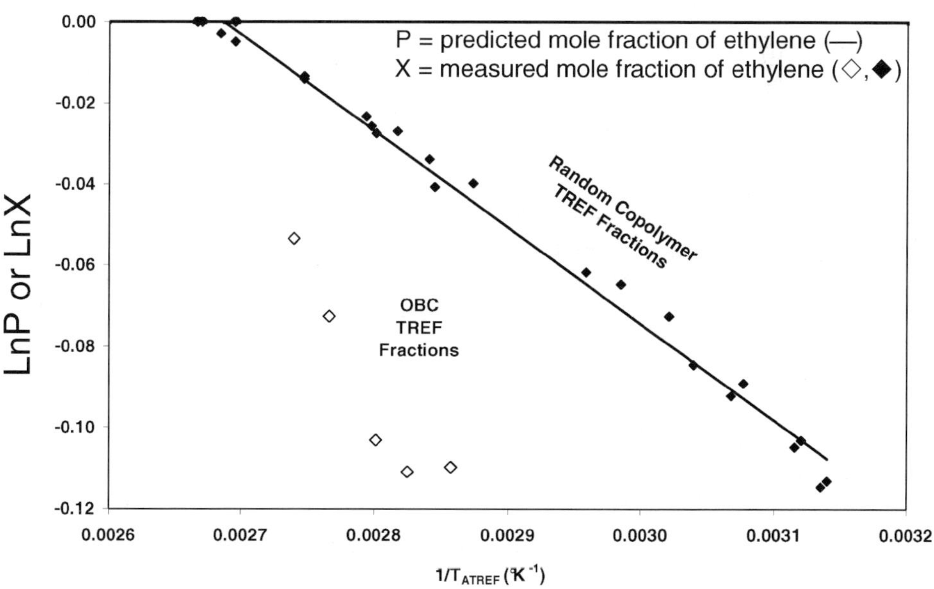

Figure 8.
OBC TREF Fractions (◇) have higher comonomer content than random equivalents (◆).

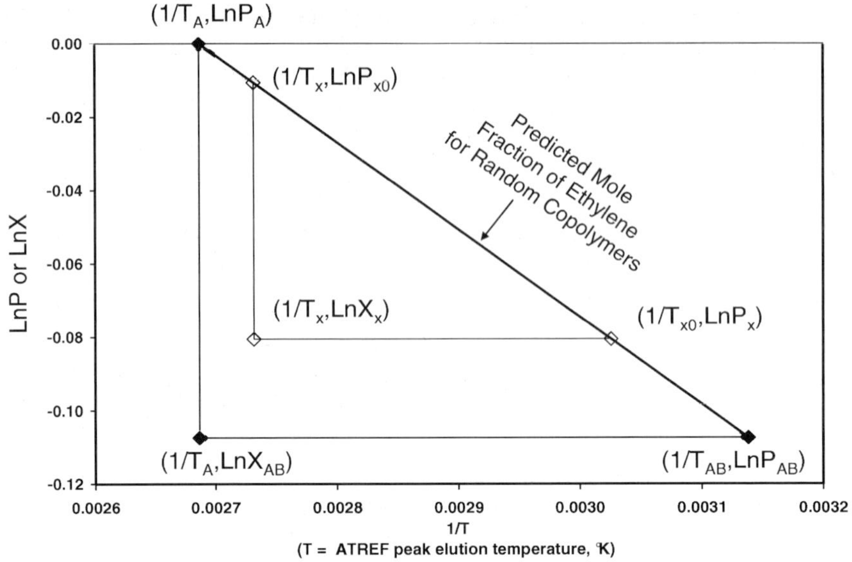

Figure 9.
Graphical definitions for the "Block Index" methodology.

to the calculated random copolymer equivalent ATREF elution temperature based on the measured P_{AB}. From the measured ATREF elution temperature, T_X, the corresponding random ethylene mole fraction, P_{XO}, can also be calculated. Similarly, from the measured NMR composition, X_X, the corresponding random elution temperature, T_{XO}, can be calculated. The square of the block index (BI) is defined to be the ratio of the area of the (T_X, X_x) triangle and the (T_A, X_{AB}) triangle, as described in Equations (2) and (3). Since the right triangles are similar, the ratio of areas is also the squared ratio of the distances from (T_A, X_{AB}) and (T_X, X_X) to the random line. In addition, the similarity of the right triangles means the ratio of the lengths of either of the corresponding sides can be used instead of the areas.

$$\text{Block Index} = \frac{1/T_x - 1/T_{x0}}{1/T_A - 1/T_{AB}} \quad (2)$$

$$\text{Block Index} = \frac{LnX_x - LnP_{x0}}{LnP_{AB} - LnP_A} \quad (3)$$

The application of this calculation to TREF fractions from the OBC is shown in Figure 10 and Tables 7–8. As shown previously, the TREF fractions from the OBC lie well below the line established for random copolymers on the ln P versus $1/T_{\text{ATREF}}$ plot in Figure 8. Using the definition for T_A hard segment temperature which was set to 372.15°K (99 °C) and the linear regression parameters for the random calibration line (Table 7), the block indices were computed from the fraction's ATREF elution temperature, and molar fraction of ethylene measured from NMR. As shown, the BI values range from 0.36 to 0.69, where a higher BI value indicates a greater deviation from the expected random behavior. In contrast, for the TREF fraction from the blend which had an ATREF elution temperature of 96 °C and 0.6 mol% octene, the calculated BI value was zero; thus indicating that the fraction is random in nature. This last result was expected since no chain shuttling agent was used to make this polymer.

Now that the method to calculate the BI value has been established, to further quantify these parameters, an average BI and BI breadth can be calculated using the amount of polymer recovered for each

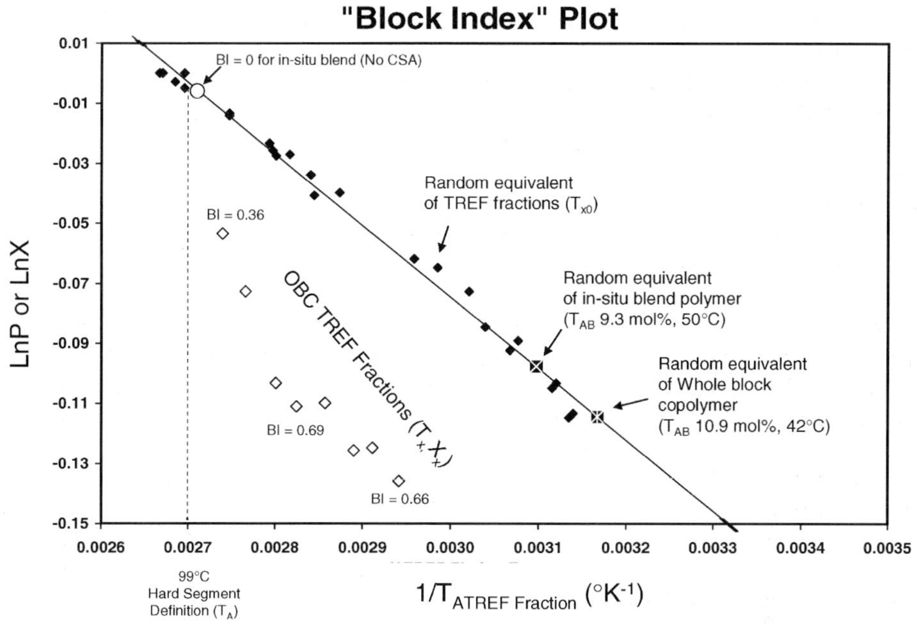

Figure 10.
Block Index results for TREF fractions of an OBC.

fraction. This provides a weighted average of the block index and can be used as a value to quantify the overall deviation of the whole polymer from the expected random behavior, as shown in Equation (4). Additionally, by calculating the second moment about the mean, a breadth index can be developed that relates to the uniformity of the block indices observed, see Equation (5).

$$\text{Average BI} = \overline{BI} = \sum (w_i \cdot BI_i) \tag{4}$$

$$\text{BI Breadth} = \sqrt{\frac{\sum w_i (BI_i - \overline{BI})^2}{\frac{(N-1)\sum w_i}{N}}} \tag{5}$$

Table 8 shows the complete calculation for the OBC example. The average block index (ABI) for this OBC is estimated to be 0.531 and the block index breadth is 0.136. Table 9 shows the block index results for the in-reactor blend (produced with no CSA) and the Ziegler-Natta LLDPE and single-site LLDPE. As expected, the block index average is reported to be zero for these polymers.

Further Applications of the Block Index

As demonstrated above, the method can be used to quantify the deviation of olefin block copolymers with a crystallizable hard

Table 7.
Preliminary values needed for block index calculations.

Variable Name	Value	Explanation
tSlope	−237.8341	Slope of regression line for Log_e of the mole fraction of ethylene versus reciprocal of analytical TREF elution temperature (°K)
tIntc	0.6390	Intercept of regression line for Log_e of the mole fraction of ethylene versus reciprocal of analytical TREF elution temperature (°K)
T_A	372.15	Analytical TREF elution temperature (°K) of hard segment
P_A	1.000	Mole fraction of ethylene of hard segment
P_{AB}	0.892	Mole fraction of ethylene in whole polymer
T_{AB}	315.722	Equivalent analytical TREF elution temperature (°K) of whole polymer calculated from whole polymer ethylene content

Table 8.
Fractionation Block Index (BI) calculations.

Fraction #	Weight Recovered (g)	ATREF Elution Tempera-ture (°K)	Mole Fraction Ethylene (NMR)	Weight Fraction Recovered	Random Equivalent ATREF Tempera-ture from NMR Ethylene Weight Fraction (°K)	Random Equivalent mole fraction ethylene from ATREF Tempera-ture	Fractional Block Index based on Tempera-ture formula	Fractional Block Index based on Log_e of mole fraction formula	Weighted Fractional Block Indices	Weighted Squared Deviations about the Weighted Mean
	Array Variable Name->	T_x	X_x	w	T_{XO}	P_{XO}	fBI	fBI	w * fBI	w * (fBI - ABI)
1	3.0402	(Note 1)	0.859	0.165	(Note 1)	(Note 1)	0	0	0	(Note 1)
2	1.9435	340	0.873	0.106	307	0.941	0.659	0.659	0.070	0.0017
3	0.7455	343.5	0.883	0.041	312	0.948	0.622	0.622	0.025	0.0003
4	1.0018	346	0.882	0.054	311	0.953	0.676	0.676	0.037	0.0011
5	2.3641	350	0.896	0.128	318	0.960	0.607	0.607	0.078	0.0007
6	4.1382	354	0.895	0.225	317	0.968	0.684	0.684	0.154	0.0052
7	3.5981	357	0.902	0.195	320	0.973	0.665	0.665	0.130	0.0035
8	1.2280	361.5	0.930	0.067	334	0.981	0.470	0.470	0.031	0.0003
9	0.3639	365	0.948	0.020	343	0.987	0.357	0.357	0.007	0.0006
Totals:	18.4233	Total Weight		1.000	Normalization check		Weighted Sums		0.531	0.0135

Final Calculations

Weighted Average Block Index (ABI)	0.531
Partial sum of weights with fBI > 0 (See Note 2 above)	0.835
Square root of sum of weighted squared deviations about the weighted mean	0.136

Note 1: Fraction #1 does not crystallize in the analytical ATREF and is assigned fBI = 0.

Note 2: The weighted squared deviations about the weighted mean use only fBI >0.

block or segment from the expected comonomer content versus elution temperature relationship for statistically random copolymers. The most perfect block distribution (ABI equal to unity) would correspond to a whole polymer with a single eluting fraction at the point (T_A, P_{AB}). A polymer that elutes at this point would preserve the ethylene segment distribution in the "hard segment", yet contain all the available comonomer (presumably in runs that are nearly identical to those produced by the soft segment catalyst). In most cases the "soft segment" will not crystallize in the ATREF (or preparative TREF). For the examples shown, the average block index was calculated for the crystallizable polymer that elutes. To include the non-crystallizing polymer contained in the purge peak (<30 °C), block index Equation (2) and (3) can still be applied. However special consideration is required to obtain lower temperature resolution or extend the calibration using DSC measurements of the purge fraction. In the case where more resolution is required, it is possible to extend the comonomer versus temperature calibration to lower temperatures by operating the ATREF to subambient conditions using a solvent such as ortho-dichlorobenzene.

Overall, the proposed "Block Index" methodology is useful to quantify the deviation of an olefin block copolymer from a polymer's expected monomer composition versus the ATREF elution temperature. If differences are observed from this relationship, it can be concluded that the intrachain distribution of comonomer

Table 9.
Block Index summary of OBC and random copolymers.

Example	Type	Chain Shuttling Agent?	Average Block Index	BI Breadth
In-reactor Blend	Blend	No CSA	0	0
OBC	Chain Shuttling	CSA	0.531	0.136
Random References				
ATTANE 4203	Ziegler-Natta	NA	0	–
AFFINITY PL1880	Single-site	NA	0	–

within the polymer is not random and that the blocks or segments within the polymer are blocked. Block indices can have a range of values between zero and unity, with unity being the largest possible deviation from random based on the hard block composition.

Conclusions

A unique structural feature of olefin based block copolymers is that the intra-chain distribution of comonomer is segmented (statistically non-random). Fractionating an olefin block copolymer by preparative temperature rising elution fractionation, TREF, results in fractions that have much higher comonomer content than comparable fractions of a random copolymer collected at an equivalent TREF elution temperature.

We have developed a "block index" methodology which quantifies the deviation from the expected monomer composition versus the analytical temperature rising elution fractionation, ATREF, elution temperature. When interpreted properly, this index indicates the degree to which the intra-chain comonomer distribution is segmented or blocked. The unique crystallization behavior of block copolymers determine the magnitude of the block index values because the highly crystalline segments along an otherwise non-crystalline chain tend to dominate the ATREF (and DSC) temperature distributions.

[1] Patent application publications WO2005090427, WO2005090426, WO2005090425.
[2] D. J. Arriola, E. M. Carnahan, P. D. Hustad, R. L. Kuhlman, T. T. Wenzel, *Science* **2006**, *5*, 714.
[3] S. Karande, SPE Annual Technical Conference 2006, Charlotte, May 7–11, 2006.
[4] Z. Zhou, R. Kummerle, X. Qiu, D. Redwine, R. Cong, A. Taha, D. Baugh, B. Winniford, *J. Magnetic Resonance* **2007**, *187*, 225.
[5] H. P. Wang, D. Khariwala, W. Cheung, S. Chum, A. Hiltner, E. Baer, *Macromolecules* **2007**, *40*, 2852.
[6] L. Wild, Temperature Rising Elution Fractionation, *Adv. Pol. Sci.* **1990**, *98*, 1.
[7] R. Brull, H. Pasch, H. G. Raubenheimer, R. Sanderson, A. J. van Reenen, W. M. Wahner, *Macromol. Chem. Phys.* **2001**, *202*, 1281.
[8] P. J. Flory, *Trans. Faraday Soc.* **1955**, *51*, 848.

DOI: 10.1002/masy.20070051108

A Mathematical Model for the Kinetics of Crystallization in Crystaf

Siripon Anantawaraskul,[*1] *João B.P. Soares,*[2] *Preechathorn Jirachaithorn*[1]

Summary: A series of ethylene homopolymers and ethylene/1-hexene copolymers with different molecular weight distributions (MWD) and chemical composition distributions (CCD) was analyzed by crystallization analysis fractionation (Crystaf) at several cooling rates to investigate the effect of MWD, CCD, and cooling rate on their Crystaf profiles. Using these results, we developed a mathematical model for Crystaf that considers crystallization kinetic effects ignored in all previous Crystaf models and can fit our experimental profiles very well.

Keywords: chemical composition distribution; crystallization analysis fractionation (Crystaf); modeling; molecular weight distribution; polyethylene

Introduction

Consisting of simple monomeric units, polyethylene and its analog, hydrogenated polybutadiene, have been viewed as simple model polymers. This is partly true for some specially synthesized samples having chain microstructures that allow us to draw several conclusions on the effect of chain microstructure and topology on physical properties. However, the notion that we have thoroughly understood this polymer is far from the true. In fact, the chain microstructure and topology of commercial polyethylene are far from simple, but rather extremely complex.

Average microstructural properties such as number average molecular weight and average comonomer content are insufficient to describe the physical properties of such complex polymers. Details on both intra- and intermolecular heterogeneity (e.g., molecular weight distribution, chemical composition distribution, sequence length distribution, and long chain branching level) are often required to fully describe commercial polyethylene. Despite the fact that understanding chain microstructure and structure-property relationships is crucial for developing new generations of polymers, establishing reliable structure-property relationships remain a challenging task for polymer engineers.

Crystallization analysis fractionation (Crystaf) is an important polyolefin characterization technique because it can analyze crystallizability distribution of semi-crystalline polymers and this distribution can be used to infer the molecular weight distribution (MWD) of homopolymers, the chemical composition distribution (CCD) of binary copolymers, and the tacticity distribution (TD) of stereoregular polymers.[1–3] Quantitative distributions can be obtained using a calibration curve, a relationship between chain crystallizability and chain microstructure for each case.

Crystaf involves the non-isothermal crystallization of polymer chains from a dilute solution. During crystallization, the concentration of polymer remaining in solution is monitored as a function of crystallization temperature (integral Crystaf distribution). The first derivative of the integral Crystaf distribution (differential Crystaf distribution) represents the weight fraction of polymer that crystallizes at each temperature.

[1] Department of Chemical Engineering, Kasetsart University, 50 Phaholyothin Rd., Jatujak, Bangkok, Thailand 10900
Tel: 662-942-8555 ext. 1231, Fax: 662-561-4621
E-mail: fengsia@ku.ac.th

[2] Department of Chemical Engineering, University of Waterloo, Waterloo, Ontario, Canada N2L 3G1

Table 1.
Properties of polyethylene and poly(ethylene-*co*-hexene) samples.

Sample (Trade Name)	Polymer Type	Number average molecular weight (M_N)	Polydispersity Index (PDI)	Mol percent of 1-hexene
PE8	Ethylene homopolymer	7,900	3.20	0
PE16 (SRM1475)	Ethylene homopolymer	15,400	3.51	0
PE32 (SRM1483)	Ethylene homopolymer	31,600	1.31	0
PE48	Ethylene homopolymer	47,900	2.15	0
EH06	Ethylene/1-hexene copolymers	36,100	2.5	0.68
EH15	Ethylene/1-hexene copolymers	35,200	2.35	1.51
EH31	Ethylene/1-hexene copolymers	34,300	2.18	3.14

Several attempts have been made to model Crystaf profiles. The models proposed in the literature can be divided into two groups: models based on Stockmayer's bivariate distribution,[4–6] and models based on Monte Carlo simulation.[7–9] Although these models can describe Crystaf profiles for a certain set of samples and fractionation conditions, they suffer from a major conceptual flaw because they assume that the fractionation occurs at, or near to, thermodynamic equilibrium. We have recently shown that this is not the case and that crystallization kinetic effects influence Crystaf analysis.[10] For a typical operation condition (a cooling rate of 0.1 °C/min), Crystaf is far from thermodynamic equilibrium, because using slower cooling rates broaden Crystaf peaks and shift them to higher crystallization temperatures.

In this paper, we propose a new semi-empirical mathematical model that accounts for the effect of crystallization kinetics during Crystaf analysis. The model was validated by fitting experimental Crystaf profiles measured at several cooling rates for a series of ethylene homopolymers and ethylene/1-hexene copolymers. Good agreement between the experimental data and the model was obtained for all the samples investigated.

Experimental Part

Materials

Seven polyethylene samples (four ethylene homopolymers and three ethylene/1-hexene copolymers) were used in this investigation. Table 1 summarizes some average properties of these samples. Two of the samples (PE16 and PE32, with trade names SRM1475 and SRM1483, respectively) were purchased from the National Institute of Standards and Technology (NIST, USA). The other polyethylene samples were synthesized in the olefin polymerization laboratory at University of Waterloo in a 300 mL Parr autoclave reactor operated in semi-batch mode. A detailed description of the polymerization procedure can be found in a previous publication.[11]

Crystaf

Crystaf analysis was performed using a Crystaf model 200 manufactured by PolymerChar S.A. (Valencia, Spain). The polymer was dissolved in 1,2,4 trichlorobenzene (TCB) in a 60 mL, stirred crystallization vessel at a concentration of 0.1 mg/mL. The polymer solution was held at 160 °C for 60 min to ensure the complete dissolution of the polymer. Then, the temperature of the solution was decreased to 110 °C and kept at that temperature for 45 min for stabilization before starting the fractionation. During analysis, the temperature of the crystallization vessel was reduced to 30 °C under constant cooling rates (0.02–1.0 °C/min). The decrease in polymer concentration in TCB solution with temperature was monitored using an in-line infrared detector. The amount of polymer crystallized at each temperature was obtained by numerical differentiation.

Crystaf Model

Model Formulation for Homopolymers

For isothermal polymer crystallization, the relationship between crystallinity, $X(t)$, and time, t, can be described using the Avrami equation,

$$X(t) = 1 - \exp(-k \cdot t^n) \quad (1)$$

where n and k are the Avrami constants. This equation can be used to describe crystallization from polymer melts and from polymer solutions.[12–14] The fractionation in Crystaf is not isothermal; however, for a group of homopolymers with similar chain microstructures, the range of crystallization temperatures during Crystaf analysis is very narrow. Therefore, we first make the assumption that both Avrami parameters can be considered constant for each homopolymer at each condition.

The Avrami exponent, n, is known to be constant over a range of temperatures, so it should be a constant over a range of cooling rates as well. However, the Avrami parameter k depends greatly on the crystallization temperature and, therefore, on cooling rate. The parameter k used in the model for homopolymers should be considered an effective or apparent parameter (i.e., an average value measured over a range of temperatures) at each cooling rate. We will show that, despite of this simplification, we can still use the model to describe the Crystaf profiles very well for the polyethylene samples studied in this investigation.

To use Equation (1), we must establish a relationship between the crystallization temperature, T_C, and the crystallization time, t. Generally, a slow, constant cooling rate, CR, is used during Crystaf analysis. Therefore, the relationship between crystallization temperature and time can be simply written as:

$$\frac{dT_C}{dt} = -CR \quad (2)$$

At the onset of crystallization ($t = 0$), the crystallization temperature should be equal to the dissolution temperature, T_d. For the case of homopolymers, T_d is a function of kinetic chain length, r. The modified Gibbs-Thomson equation introduced by Beigzadeh et al.[7] was used for this purpose:

$$T_d(r) = T_d^o\left[\frac{r-\alpha}{r}\right] - T_S \quad (3)$$

In Equation (3), T_d^o is the equilibrium dissolution temperature of a chain with infinite length (this parameter depends on solvent and polymer type), T_S is a supercooling temperature (to account for supercooling during Crystaf analysis), and α is a constant that is inversely proportional to the enthalpy of fusion. To reduce the number of parameters in the model, Equation (3) is rearranged as follows,

$$T_d(r) = A - \frac{B}{r} \quad (4)$$

where

$$A = T_d^o - T_S \text{ and } B = T_d^o \times \alpha \quad (5)$$

Since both T_d^o and T_S are essentially constant for a given polymer/solvent combination, we expect the value of the parameter A to remain constant. Similarly, because α is a constant that is inversely proportional to the enthalpy of fusion, the parameter B should also remain constant for all cooling rates.

We also have to consider the difference between the temperature measured in the Crystaf oven and the temperature inside the crystallization vessel (temperature lag, T_l). For our Crystaf instrument, the empirical relation between the temperature lag and cooling rate was reported in our previous publication:[10]

$$T_l = 5.02 \times CR - 0.05 \quad (6)$$

This empirical equation was established for a cooling rate range of 0.02–1 °C/min, which covers the experimental conditions studied in the present investigation.

Considering the temperature lag in the system, the initial condition for Equation (2) is $T_C(0) = T_d(r) - T_l$. We must integrate Equation (2) with this initial condition to obtain the relationship between crystallization temperature and time:

$$t = \frac{(T_d(r) - T_l) - T_C}{CR} \quad (7)$$

We obtain our final equation relating the degree of crystallinity as a function of chain length, cooling rate, and crystallization temperature for homopolymers by combining Equation (1), (4), and (7):

$$X(r,T_C)=\begin{cases}0 & ,\ T_C \geq T_d(r)-T_l \\ 1-\exp\left\{-k\left[\frac{(A-B/r-T_l)-T_C}{CR}\right]^n\right\} & ,\ T_C < T_d(r)-T_l\end{cases} \quad (8)$$

Since crystallization takes place only when the temperature is lower than $T_d - T_l$, the crystallinity is set to 0 when the temperature is higher than $T_d - T_l$. Because crystallization takes place from a dilute solution, an increase in crystallinity is related to a decrease in the weight fraction of polymer remaining in solution, $C(T_C)$. We propose a very simple equation to describe this relationship (the integral Crystaf profile):

$$C_{model}(T_C) = \sum_{r=1}^{\infty} m(r)\cdot(1-X(r,T_C)) \quad (9)$$

where $m(r)$ is the weight fraction of chains with kinetic chain length of r. The differential Crystaf profile can be obtained by differentiation of Equation (9) with respect to the crystallization temperature, i.e. by calculating $dC_{model}(T_C)/dT_C$.

The model parameters (n, k, A, and B) were determined by minimizing the sum of the squares of differences between the simulated and experimental Crystaf profiles (Φ):

$$\Phi = \sum_{T_C=T_{initial}}^{T_{final}} \left(C_{model}(T_C) - C_{exp}(T_C)\right)^2 \quad (10)$$

Model Formulation for Binary Copolymers

The Crystaf model for homopolymers can be extended to binary copolymers such as ethylene/1-olefin copolymers. In the case of homopolymers, molecular weight is considered as the key factor influencing the chain crystallizabilities. However, in the case of binary copolymers, the longest ethylene sequence (LES),[7] which depends on both molecular weight and comonomer content, is considered to be the key factor determining chain crystallizability.[7] The theoretical expression describing the weight distribution function of LES for random copolymer, $W(LES)$, was reported in our previous publication:[8]

$$W(LES) = \frac{(1-P_a)(1-pp)}{P_a}\begin{bmatrix}F(1-pm^{LES}) \\ -F(1-pm^{LES-1})\end{bmatrix} \quad (11)$$

The function $F(K)$ and parameter P_a in Equation (11) are defined as,

$$F(K)=\frac{P_a.K}{(1-P_a.K)^2}\left[LES(1-\frac{P_a.K}{1-pm^{LES}})+\frac{P_a.K}{1-pm}\right] \quad (12)$$

$$P_a=\frac{pp(1-cp)}{1-(cp.pp)} \quad (13)$$

where pm is the monomer addition probability ($pm = pp.cp$), cp is the monomer/comonomer choice probability ($cp = 1\text{-}cpp$), cpp is the comonomer propagation probability, and pp is the propagation probability. Note that the comonomer propagation probability (cpp) is equal to the average comonomer mole fraction in the copolymer for random copolymers. The chain propagation probability (pp) can be calculated from the number average chain length (r_N) and average comonomer content with the expressions,

$$pp=\frac{r_N-1}{r_N} \quad (14)$$

$$r_N=\frac{M_N}{M_{CO}\times cpp + M_{MO}(1-cpp)} \quad (15)$$

where M_N is the number average molecular weight, M_{CO} is the molecular weight of the comonomer (84 g/mol for 1-hexene), and

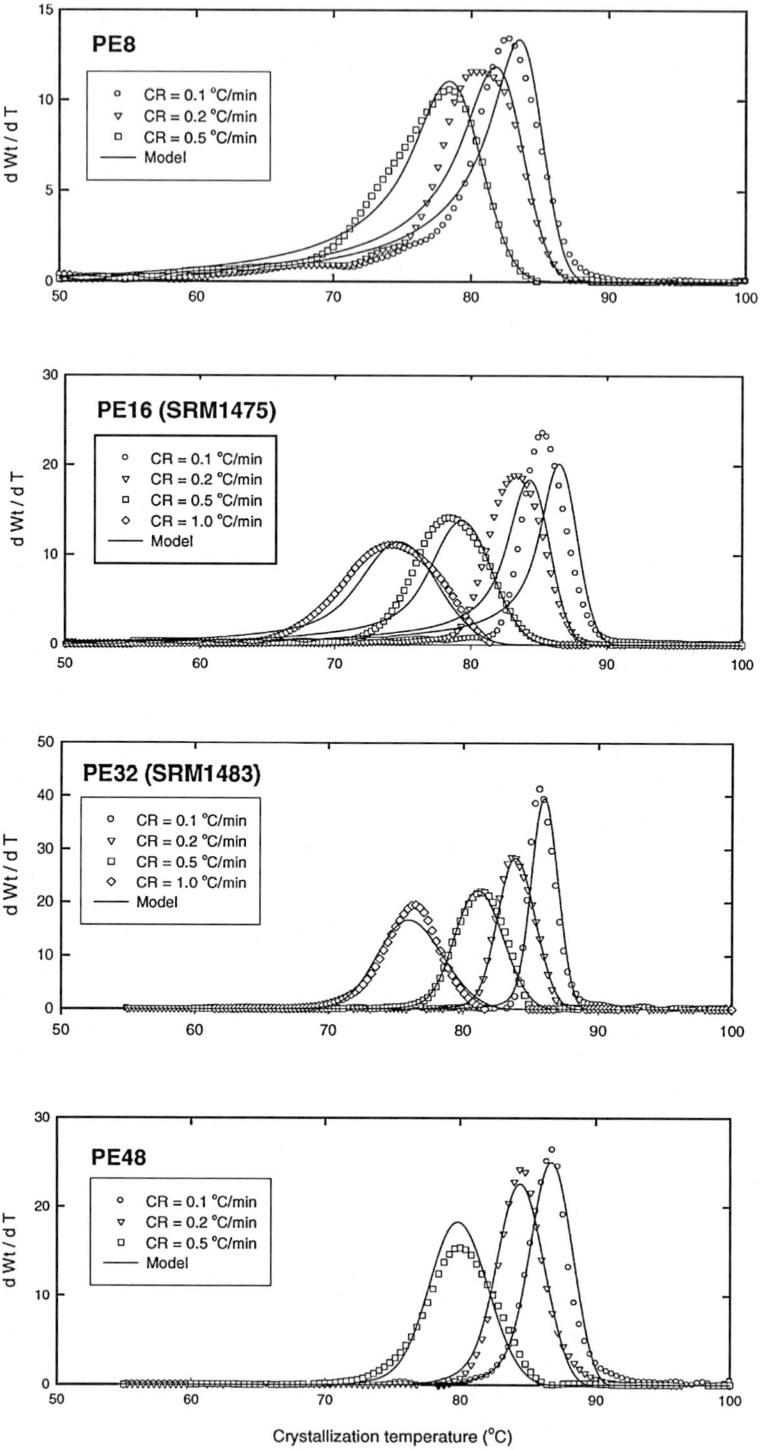

Figure 1.
Comparison between experimental and modeled Crystaf profiles for polyethylene measured at several cooling rates.

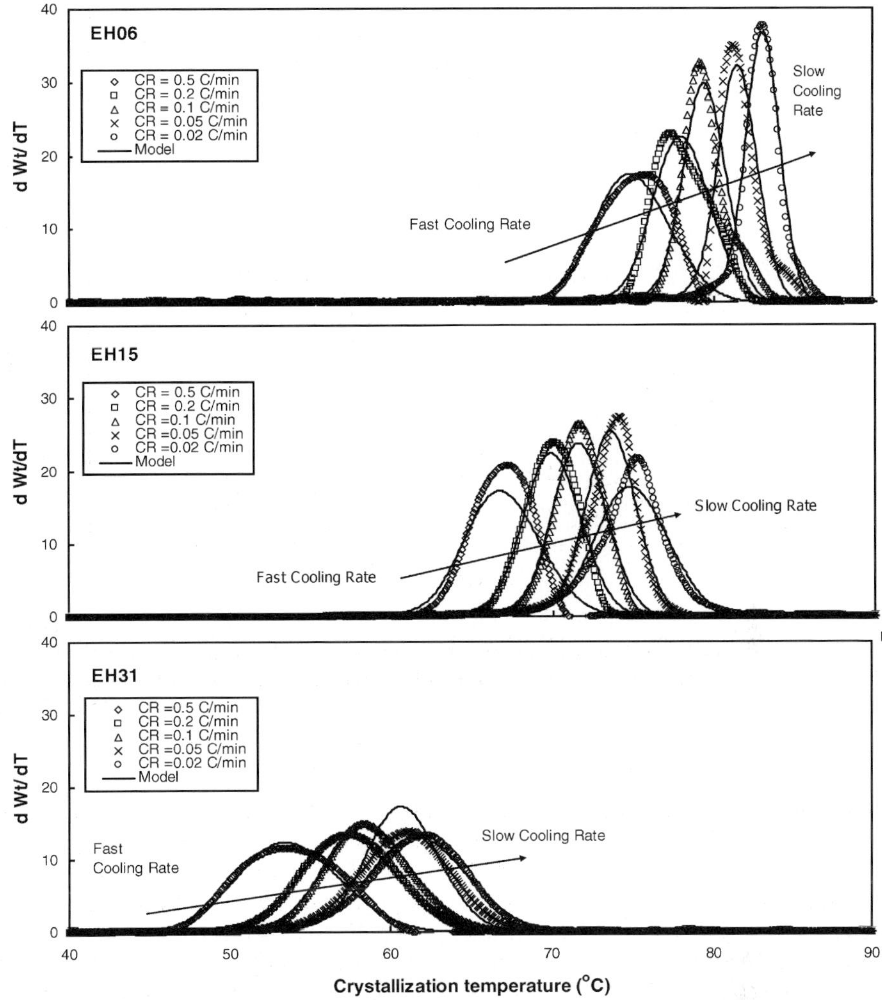

Figure 2.
Comparison between experimental and modeled Crystaf profiles for ethylene/1-hexene copolymers measured at several cooling rates.

M_{MO} is the molecular weight of the monomer (28 g/mol for ethylene).

Following the assumption that the longest ethylene sequence (LES) is the key factor governing chain crystallizability, instead of molecular weight as for homopolymers, we can rewrite Equation (3) to (8) substituting r by LES. Note that the parameters A and B for binary copolymers in Equation (5) are no longer constant as the equilibrium dissolution temperature is a function of comonomer content.

Taking the information from the weight distribution function of LES, W(LES), calculated from Equation (11) into account, the simulated Crystaf profile of copolymers can be calculated as:

$$C_{model}(T_C) = \sum_{r=1}^{\infty} (W(LES) \cdot (1 - X(LES, T_C))) \tag{16}$$

Using the same approach, the model parameters (n, k, A, and B) were determined by minimizing the sum of the squares of the differences between the simulated and experimental Crystaf profiles (Φ), Equation (10).

Results and Discussion

Figure 1 and 2 compare experimental and modeled Crystaf profiles polyethylene and ethylene/1-hexene copolymers measured at several cooling rates. The proposed model adequately describes the effect of cooling rate on Crystaf profiles for all samples and correctly follows the broadening of the distributions due to crystallization kinetic effects. The model also captures well the comonomer content effect by predicting the shift of Crystaf profiles to lower crystallization temperatures and the broadening of the distributions as the comonomer content increases.

Optimum model parameters (n, k, A, and B) were obtained by minimizing the function Φ in Equation (10). Figure 3 and 4 shows the Avrami and modified Gibb-Thomson parameters estimated for the polyethylene samples at several cooling rates. As expected, the values of n, A, and B are practically independent of cooling rate (or crystallization temperature). For polyethylene, the average values for these three parameters are $n = 3.96$, $A = 90.45\,^{\circ}C$, and $B = 654.75\,^{\circ}C \cdot$ (number of repeating unit). Previous investigations on polymer crystallization from solution by Devoy et al.[12]

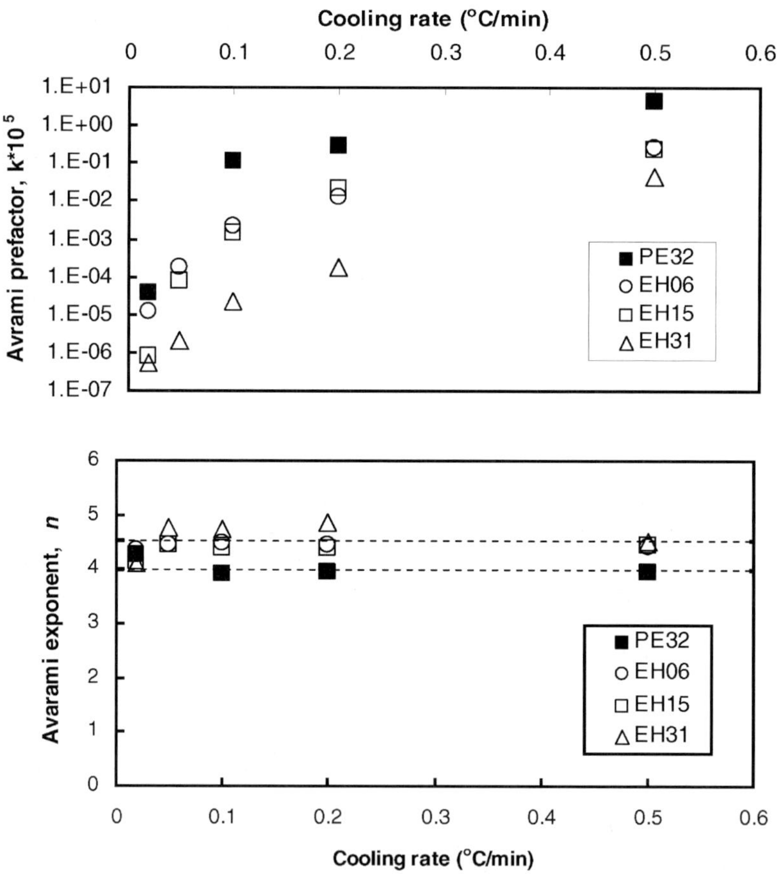

Figure 3.
Estimated Avrami parameters for the Crystaf model. The parameter *k* is an apparent or average value. (The dashed line is the average value for the estimated parameters).

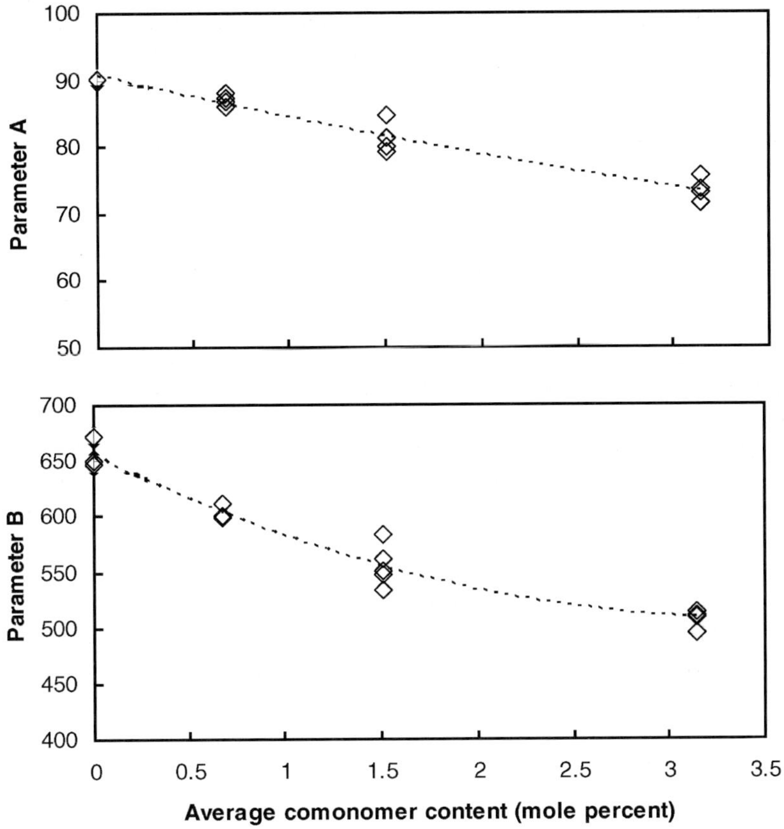

Figure 4.
Estimated Gibbs-Thomson parameters for the Crystaf model as a function of average comonomer content (The dashed lines are to aid the eye only).

and Raiande and Fatou[13–14] showed that the closest integer for parameter n for polyethylene in several solvents was 4. This agrees very well with an average n value of 3.96 obtained from our parameter estimation. Our average value of parameter A (90.45 °C) is also close to the values for the crystallization temperature of ethylene homopolymers in several solvents (85–90 °C) reported earlier by Jackson and Mandelkern.[15]

Our estimated value for the parameter k is in the range of 2×10^{-8} to 16×10^{-5} min^{-n}. To the best of our knowledge, no values have been reported for the parameter k for polyethylene in TCB, but k values for polyethylene crystallized in several other solvents are in the range of 10^{-5} to 10^{-15} min^{-n}.[12–14] It has also been reported that the parameter k increases as crystallization temperature decreases. This observation agrees well with our results since we show that k increases with increasing Crystaf cooling rates (the faster the cooling rate, the lower the crystallization temperature.)

For ethylene/1-hexene copolymers, the estimated values of the parameter n are relatively constant, as expected; however, the average value of the parameter n of ethylene/1-hexene copolymers ($n = 4.49$) is slightly higher than the one for polyethylene ($n = 3.96$). The value of parameter k was found to increase with a cooling rate as we had also observed for the polyethylene samples. Moreover, the k value of ethylene/

1-hexene copolymers decreases as comonomer fraction increases. This might be because the comonomer unit disrupts the regularity of the chain, thus lowering the chain crystallizability and rate of crystallization; similar observations have been reported in the literature.[16]

Parameters *A* and *B* were found to be relatively independent of the cooling rate used, but strongly dependent on comonomer content. The trends show that both parameters *A* and *B* decrease as comonomer content increases. This is expected and can be explained from Equation (5) because the equilibrium dissolution temperature decreases as the comonomer fraction in the copolymer increases.

In summary, we believe that our model performed quite well considering that the only input for predicting the Crystaf profile of homopolymers is molecular weight and for predicting the Crystaf profile of copolymers is molecular weight and comonomer content.

The model certainly captures the effect of cooling rate on Crystaf profiles peak positions and width. In the case of polyethylene, the model parameters were shown to be within the range of earlier reported values. In the case of ethylene/1-hexene copolymers, these parameters deviate from those found for the homopolymer, but do follow the expected trends.

Applications of this model include construction of the generic Crystaf calibration curves for data interpretation. These calibration curves based on a theoretically sound model could lead to a more efficient use of Crystaf. For example, one may perform an analysis at fast cooling rate (shorter analysis time) and use the calibration curve to help predict the Crystaf profile that would be measured at a slower cooling rate. Demonstration of this approach will be the subject of a future publication.

Conclusions

We developed a new model for Crystaf analysis of homopolymers and binary copolymers that considers the kinetics of crystallization based on the Avrami equation. The model could fit the experimental Crystaf profiles of polyethylene resins measured at a broad range of cooling rates. In addition, the model parameters are theoretically sound and agree with values previously reported in the literature for similar systems. The model also describes well how molecular weight and comonomer content affects the Crystaf profiles of polyethylene and ethylene/1-olefin copolymer samples.

[1] B. Monrabal, *J Appl Polym Sci.* **1994**, *52*, 491.
[2] S. Anantawaraskul, J. B. P. Soares, P. M. Wood-Adams, *Adv Polym Sci.* **2005**, *182*, 1.
[3] J. B. P. Soares, S. Anantawaraskul, *J Polym Sci Part B: Polym Phys.* **2005**, *43*, 1557.
[4] J. B. P. Soares, B. Monrabal, J. Nieto, J. Blanco, *Macromol. Chem. Phys.* **1998**, *199*, 1917.
[5] A. A. da Silvo Filho, J. B. P. Soares, G. B. de Galland, *Macromol. Chem. Phys.* **2000**, *201*, 1226.
[6] D. M. Sarzotti, J. B. P. Soares, A. Penlidis, *J Polym Sci Part B: Polym Phys.* **2002**, *40*, 2595.
[7] D. Beigzadeh, J. B. P. Soares, T. A. Duever, *J Appl Polym Sci.* **2001**, *80*, 2200.
[8] S. Costeux, S. Anantawaraskul, P. M. Wood-Adams, J. B. P. Soares, *Macromol Theory Simul.* **2002**, *11*, 326.
[9] S. Anantawaraskul, J. B. P. Soares, P. M. Wood-Adams, B. Monrabal, *Polymer* **2003**, *44*, 2393.
[10] S. Anantawaraskul, J. B. P. Soares, P. M. Wood-Adams, *J Polym Sci Part B: Polym Phys.* **2003**, *41*, 1762.
[11] S. Anantawaraskul, J. B. P. Soares, P. M. Wood-Adams, *Macromol. Chem. Phys.* **2004**, *205*, 771.
[12] C. Devoy, L. Mandelkern, L. Bourland, *J Polym Sci Part A-2*, **1970**, *8*, 869.
[13] E. Riande, J. M. G. Fatou, *Polymer* **1976**, *17*, 99.
[14] E. Riande, J. M. G. Fatou, *Polymer*, **1976**, *17*, 795.
[15] J. F. Jackson, L. Mandelkern, *Macromolecules* **1968**, *1*, 546.
[16] E. Perez, R. Benavente, A. Bello, J. M. Perena, D. Zucchi, M. C. Sacchi, *Polymer* **1997**, *38*, 5411.

Characterization of Ethylene-1-Hexene Copolymers Made with Supported Metallocene Catalysts: Influence of Support Type

Beatriz Paredes,[1] *João B.P. Soares,**[2] *Rafael van Grieken,*[1] *Alicia Carrero,*[1] *Inmaculada Suarez*[1]

Summary: It is known that the nature of the support, as well as the technique used to anchor the metallocene onto it, play important roles on catalytic activity and on the properties of the polymers produced with supported metallocenes. In the present work, the effect of different support types on the microstructure of ethylene/1-hexene copolymers made with supported metallocene catalysts has been investigated through the analysis of molecular weight and chemical composition distributions using high temperature gel permeation chromatography (GPC) and crystallization analysis fractionation (Crystaf). The copolymer samples obtained using commercial carriers (silica and silica-alumina) had unimodal chemical composition distributions and were used to create a linear calibration curve relating the peak crystallization temperature from Crystaf and the comonomer content as determined by ^{13}C NMR. This calibration curve is useful to determine the 1-hexene fractions for each peak in the resins showing bimodal chemical composition distributions, such as those obtained with catalysts supported on MCM-41 and SBA-15 materials. The structure and chemistry of the support used had a large influence on comonomer incorporation and the shape of the chemical composition distribution of the polymer, which suggests that the supporting process creates different types of active sites.

Keywords: crystallization analysis fractionation (Crystaf); MCM-41; SBA-15; silica-alumina; supported metallocene catalysts

Introduction

Metallocene catalysis can be considered a major breakthrough in polyolefin production technology. The discovery of metallocene/alkylaluminoxane systems allowed the synthesis of polyolefins with very different properties from those made with Ziegler-Natta catalysts.

Metallocenes combine high activity with excellent polymer microstructural control[1] but, because they are homogeneous catalysts, they produce polymer particles with ill-defined morphologies that cause reactor fouling when used in slurry reactors. On the other hand, these catalysts have been used very effectively in solution processes by several companies worldwide.

The immobilization of metallocenes on supports is a technical and economical solution to these limitations.[2] Moreover, to replace the conventional heterogeneous Ziegler–Natta catalysts used in industrial slurry and gas-phase processes with metallocene catalysts (drop-in technology), metallocene catalysts have to be immobilized on supports. It is important to find a way to attach the metallocene to the support without losing the performance of the homogeneous complex, while improving the morphology of the polymer particles.[3]

[1] Department of Chemical and Environmental Technology, ESCET, Universidad Rey Juan Carlos, 28933 Mostoles, Spain
[2] Department of Chemical Engineering, University of Waterloo, Waterloo, Ontario N2L 3G1, Canada
E-mail: jsoares@cape.uwaterloo.ca

In general, when a metallocene catalyst is supported, its activity decreases because of the significant steric hindrance around the active site caused by the large support surface, deactivation of catalytic sites, or inefficient production of active sites during the supporting process.[4] It is known that the nature of the support, as well as the technique used to anchor the metallocene onto it, play important roles on catalytic activity and on the properties of the polymers produced with supported metallocenes.[5]

There are three basic methods of supporting aluminoxane-activated single-site catalysts: (1) supporting the aluminoxane first, then reacting the aluminoxane-treated support with the metallocene; (2) supporting the metallocene first, then reacting the metallocene-treated support with the aluminoxane; and (3) contacting the aluminoxane and the metallocene in solution before supporting and then adding this solution to the support. The last method has some advantages over the other two: It maximizes the number of active centers by activating the metallocene in solution instead of carrying out the process with either the metallocene or the aluminoxane immobilized on the support, reduces the preparation time, and lowers the amount of solvent required for supporting.[6]

There are many publications describing the heterogeneization of metallocenes onto supports such as silica,[7–10] alumina,[11] magnesium chloride,[12] starch,[13] zeolites,[14] cyclodextrin,[15] and synthetic polymers.[16,17] Metallocenes can also be attached onto the inner surface of mesoporous molecular sieves.[18] The large surface areas of mesoporous molecular sieves can provide high dispersion of the metallocenes. These materials allow studying the effect of different structural variables, such as pore size, pore volume and surface areas, and of chemical properties by varying the ratio of silicon to aluminium during support synthesis.[19] These properties may have a significant influence on the catalytic behaviour of metallocenes and on the properties of the obtained polymer.

In this work, mesostructured MCM-41 materials[20] with different Si/Al ratio and SBA-15 materials[21] with different pore sizes have been used to immobilize $(nBuCp)_2ZrCl_2$/MAO. Commercial carriers such as silica and silica-alumina have also been investigated for comparison.

Metallocene catalysts are generally believed to have uniform site types even after being supported. In general, polymers produced by supported metallocenes have narrow molecular weight distributions (MWD)[22,23] with polydispersity indexes close to two or slightly higher. Some supported metallocenes may also make polyolefins with broad MWDs, which has been associated with the formation of several active site types[24] and/or mass transfer resistances during polymerization.

The effect of different support types has been studied for the copolymerization of ethylene and 1-hexene. For the case of copolymers, besides MWD determination, it is necessary to measure the chemical composition distribution (CCD) to have a more complete understanding of active site types and polymer properties. Therefore, in this study the influence of support type has been investigated through the analysis of MWD and CCD using high temperature gel permeation chromatography (GPC) and crystallization analysis fractionation (Crystaf), respectively. It will be shown that polyolefins having a narrow MWD can have a very broad and multimodal CCD, indicating the presence of more than one active site type during the polymerization.

Experimental Part

Synthesis of the Carriers and Preparation of Supported Catalysts

Commercial carriers silica and silica-alumina from W.R. Grace were heat-treated at 400 °C and at 200 °C, respectively, for 5 hours before use.

The MCM-41 supports were synthesized in basic medium at room temperature[20] using hexadecyltrimethylammonium bromide (C_{16}TABr, 99 wt%, Aldrich) as surfactant, water, dimethylamine (DMA, 40 wt%, Aldrich), tetraethyl ortosilicate (TEOS, 98 wt%, Aldrich) as the silica source, and aluminium (molar ratio of Si/Al = 15, 60 and infinite) was incorporated during synthesis using aluminium isopropoxide (AIP, 99 wt%, Aldrich) as the aluminium source. The surfactant was first dissolved in water and DMA. The mixture of TEOS and AIP was added to this solution to form the MCM-41 supports. The final mixture was subjected to an ageing procedure at 105 °C for 48 hours. Finally, the resulting material was filtered and calcined at 550 °C for 5 hours in static conditions.

The SBA-15 supports were prepared according to a direct synthesis procedure previously published in the literature[21,25–27] in acidic medium at 40 °C using pluronic $EO_{20}PO_7EO_{20}$ (P123, Aldrich) as surfactant, hydrochloric acid (HCl, 35 wt%, Sharlau) and tetraethyl ortosilicate (TEOS, 98 wt%, Aldrich) as the silica source. The surfactant was first dissolved in a solution of HCl (pH = 1.5), then TEOS was added and the mixture was kept at 40 °C for 20 h. In order to make silica SBA-15 supports with different pore sizes 1,3,5-trimethylbenzene (TMB, 98 %wt, Aldrich) and n-decane (98 wt%, Fluka) were used as swelling agents. After crystallization, they were aged at 105 °C for 24 hours. Finally, the resulting material was filtered and calcined at 500 °C for 5 hours under dry air flow.

Heterogeneous catalysts were prepared by impregnating commercial and synthesized supports with a mixture ($Al_{(MAO)}$/Zr = 170 mol/mol) of methylaluminoxane (MAO 30 wt% in toluene, Witco) and a solution of bis(butylcyclopentadienyl) zirconium dichloride ($(nBuCp)_2ZrCl_2$, Crompton) in dry toluene (99 wt%, Scharlab) under inert nitrogen atmosphere using Schlenk techniques and a glove box. The ratio between the volume of impregnating solution and support pore volume was three. The reaction was performed at room temperature, in a stirred vessel during 3 hours. The supported catalyst was dried under nitrogen flow and stored in a glove-box.

Characterization of Supports and Catalysts

Nitrogen adsorption-desorption isotherms at 77 K were obtained with a Micromeritics Tristar 2050 apparatus. The samples were previously out-gassed under vacuum at 200 °C for 2 hours. Surface areas were calculated by means of the BET equation whereas mean pore size was obtained from the maximum of BJH pore size distribution.

Polymerization and Polymer Characterization

Polymerizations were performed at 70 °C in a 1 litter stirred-glass reactor. The ethylene flow rate was followed by a mass-flow indicator in order to keep the reactor pressure at 5 bar during the polymerization. Different amounts of 1-hexene were injected into the reactor with a syringe at the beginning of the polymerization. In these reactions, tri-isobutylamunimum (TIBA, 30 wt% in heptane, Witco) was added as scavenger in an $Al_{(TIBA)}$/Zr molar ratio of 400. After 30 minutes, the polymerization was stopped by depressurization and quenched by addition of acidified (HCl) methanol. The polymer obtained was separated by filtration and dried under atmospheric pressure at 70 °C.

Molecular weight averages and distributions were determined with a Waters ALLIANCE GPCV 2000 gel permeation chromatograph (GPC) equipped with a refractometer, a viscosimeter and three Styragel HT type columns (HT3, HT4 and HT6) with exclusion limit 1×10^7 for polystyrene. 1,2,4-Trichlorobenzene was used as solvent, at a flow rate of 1 cm^3min^{-1}. The analyses were performed at 145 °C. The columns were calibrated with standard narrow molar mass distribution polystyrenes and with linear low density polyethylenes and polypropylenes.

Polymer melting points (T_m), crystallization temperatures (T_c) and crystallinities were determined in a METTLER TOLEDO DSC822 differential scanning calorimeter (DSC), using a heating rate of 10 °Cmin^{-1} in the temperature range 23–160 °C. The heating cycle was performed twice, but only the results of the second

Table 1.
Physical properties of the supports.

Support	SiO_2	SiO_2-Al_2O_3	MCM-41 Si/Al			SBA-15 swelling agent		
			15	60	infinite	none	n-decane	TMB
Average pore diameter (nm)[a)]	28.5	18.2	2.1	2.4	2.5	8.8	11.3	22.7
BET surface area (m^2/g)	285	370	837	1015	1129	628	608	588
Pore volume (cm^3/g)	1.55	1.29	0.58	0.77	0.77	1.16	0.78	1.73

a) Determined from the maximum of BJH pore size distribution.

scan are reported, because the former is influenced by the mechanical and thermal history of the samples.

Chemical composition distributions were measured by Crystaf (Polymer Char) using 1,2,4-trichlorobenzene as a solvent. Dissolution took place at 160 °C for 90 min followed by equilibration at 95 °C for 45 min. The crystallization rates were 0.1 °C/min from 95 to 19 °C. A two channel infrared detector was used to measure the concentration of polymer in the solution during crystallization.

A BRUKER AC300 spectrometer at 75 MHz was used to characterize the copolymers by ^{13}C NMR measurement and determine their 1-hexene molar fraction. The samples were added into sample tube (a diameter of 5 mm) with the mixture of 1,2,4-trichlorobenzene and 1,1,2,2-tetrachloroethylane-d_2 in a concentration of 10 vol%. Each sample was measured with 10 s pulse repetition.

Results and Discussion

The main physical properties of the commercial and synthesized materials are summarized in Table 1. The nitrogen adsorption–desorption isotherms at 77 K (not shown) belonged to type IV of the I.U.P.A.C classification corresponding to mesoporous materials. As it can been observed in the table, three MCM-41 materials with similar pore diameter and different silicon to aluminium molar ratio were obtained, and three SBA-15 carriers with different pore size, as a function of the swelling agent employed in the synthesis, were prepared. The effect of the structure and chemical properties of support material on the ethylene-1-hexene copolymerization has been investigated using these materials as supports of the catalytic system $(nBuCp)_2ZrCl_2$/MAO.

All the supported catalysts had good activities for polymerization, with activities varying from 0.46 to $9.91 \cdot 10^6$ g PE/mol Zr · h · bar, observing a positive comonomer effect with a different maximum value depending on the support employed.

Molecular weight averages of polymers made with $(nBuCp)_2ZrCl_2$/MAO supported on the commercial carriers were not affected significantly when 1-hexene was added (Table 2). The polydispersity index

Table 2.
Characterization of poly(ethylene-*co*-1-hexene) samples made with $(nBuCp)_2ZrCl_2$/MAO supported on commercial carriers.

Support	mL 1-hexene	0	5	15	25	35	50	60	75
SiO_2	$\overline{M}_n$(g/mol)	81061	53853	55415	58336	64847	67197	73903	75450
	$\overline{M}_w/\overline{M}_n$	2.91	3.06	2.76	2.58	2.59	2.58	2.79	2.91
	T_m (°C)	133	122	113	108	104	95	94	91
	T_c (°C)	116	112	98	98	91	86	86	73
	Crystallinity (%)	64	51	40	37	22	27	25	19
SiO_2-Al_2O_3	$\overline{M}_n$ (g/mol)	70613	48307	44687	47541	78729	67887	77617	52250
	$\overline{M}_w/\overline{M}_n$	2.89	2.70	2.63	2.78	2.62	2.87	3.09	3.28
	T_m (°C)	132	122	115	109	104	96	104	94
	T_c (°C)	117	111	102	95	87	78	85	79
	Crystallinity (%)	64	51	43	36	30	24	28	24

$(\overline{M}_w/\overline{M}_n)$ varied from 2.58 to 3.28, higher than the value of 2.0 expected for homogeneous single-site catalysts, suggesting some heterogeneity in the nature of the active sites.[28] Melting and crystallization temperatures decreased with increasing comonomer content, as expected.

Figure 1 shows the Crystaf profiles of the polymers produced with $(nBuCp)_2ZrCl_2$/MAO supported on SiO_2 and SiO_2-Al_2O_3. Crystallization temperatures are correlated to the comonomer content.[29] As all of them have narrow, unimodal CCDs, they are useful to create a calibration curve since it covers a broad range of comonomer incorporation.[30] The linear relationship between Crystaf peak temperature, T_{peak}, and the molar fraction of 1-hexene measured by ^{13}C NMR is illustrated in Figure 2.

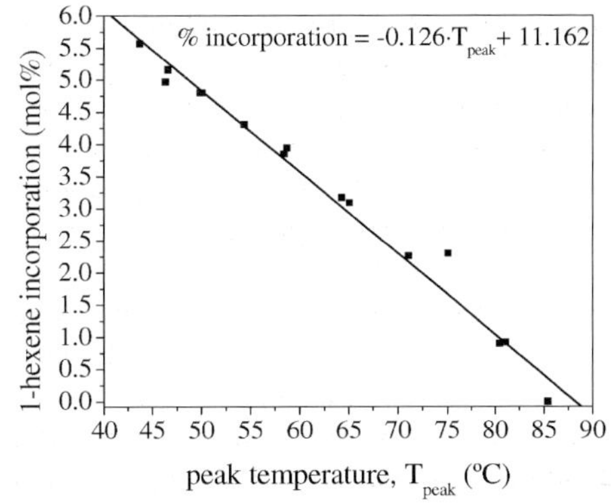

Figure 2.
Crystaf calibration curve for poly(ethylene-*co*-1-hexene) made with $(nBuCp)_2ZrCl_2$/MAO supported on commercial carriers.

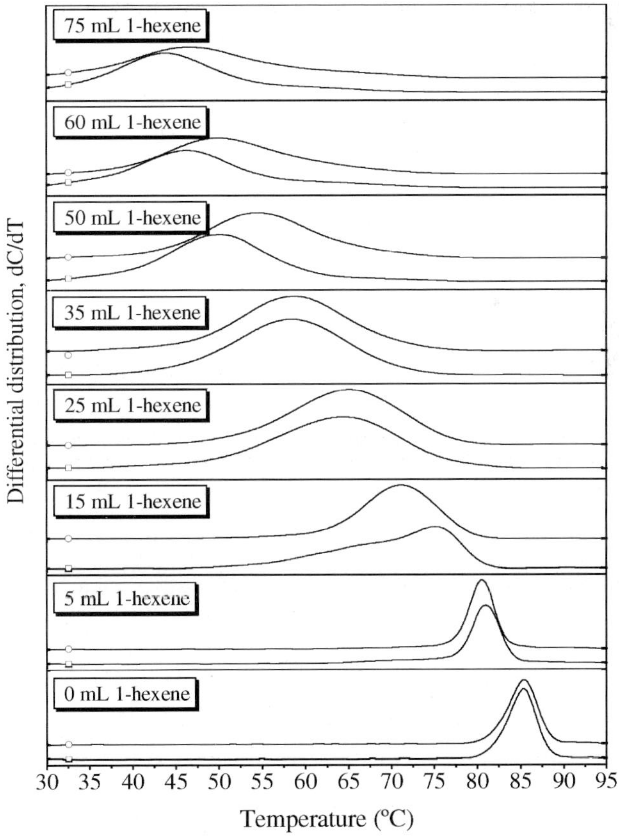

Figure 1.
Crystaf profiles of poly(ethylene-*co*-1-hexene) made with $(nBuCp)_2ZrCl_2$/MAO supported on commercial carriers: □ SiO_2 and ○ SiO_2-Al_2O_3.

Several investigations have pointed out that the formation and stabilization of zirconium cations (Zr^+) may be easier when support materials have surface acidic properties.[3,31] In this work, we have studied this phenomenon through the use of synthesized MCM-41 materials with different silicon to aluminium molar ratios. Table 3 shows that the acidity of the support does not have a significant influence on the properties of the copolymers, apart from those related with the incorporation of higher amounts of the α-olefin. When MCM-41 is used, the MWD continues to be narrow, but the CCD broadens significantly with increasing 1-hexene content and even becomes bimodal for certain 1-hexene concentrations, as shown in Figure 3. This may be related to the presence of two catalyst site types on the surface of the support. Therefore, the Si/Al ratio can be used to control the shape of the CCD. Interestingly, lower temperature peaks become more prominent with decreasing Si/Al ratios.

Another important property affecting the behaviour of support materials for metallocenes is pore size, since the support has to be able to anchor the catalyst precursor, the cocatalyst, and permit easy access to the active sites for the monomers.[18] The geometrical constraints of the parallel hexagonal channel structure of the SBA-15 materials may affect the pattern of monomer insertion and chain growth process, which offers a way to control the polymer chain structure and morphology during the polymerization.[32] To study this effect, we have synthesized three SBA-15 materials with different pore sizes using several swelling agents, as described above.

Table 4 summarizes the properties of the poly(ethylene-*co*-1-hexene) made with $(nBuCp)_2ZrCl_2$/MAO supported on SBA-15 materials. The MWD is still narrow, but the CCD can also become broad and bimodal, as depicted in Figure 4. Larger pore sizes seem to favour 1-hexene incorporation slightly, which may be related in part to intraparticle mass transfer resistances.

Finally, we have calculated the overall mol% of 1-hexene for the copolymer samples obtained with $(nBuCp)_2ZrCl_2$/MAO supported on both MCM-41 and SBA-15 using the calibration curve shown in Figure 2. These results are reported in Table 5, together with overall 1-hexene mol% measured with ^{13}C NMR. The good agreement between the two techniques indicates that our calibration curve also works well for the bimodal CCD copolymers made with $(nBuCp)_2ZrCl_2$/MAO supported on both MCM-41 and SBA-15.

Table 3.
Characterization of the poly(ethylene-*co*-1-hexene) made with $(nBuCp)_2ZrCl_2$/MAO immobilized on supports synthesized with different Si/Al ratios.

Support	mL 1-hexene	0	5	25	50	75
MCM-41-15	$\overline{M}_n$ (g/mol)	62358	37521	44796	63216	54288
	$\overline{M}_w/\overline{M}_n$	2.95	2.85	2.94	3.3	3.58
	T_m (°C)	136	124	109	103	100
	T_c (°C)	113	110	95	95	82
	Crystallinity (%)	66	54	36	32	26
MCM-41-60	$\overline{M}_n$(g/mol)	60915	38927	46271	60234	52746
	$\overline{M}_w/\overline{M}_n$	2.9	2.7	2.75	3.09	3.85
	T_m (°C)	133	123	110	110	110
	T_c (°C)	116	110	91	94	86
	Crystallinity (%)	62	53	36	32	31
MCM-41-infinite	$\overline{M}_n$(g/mol)	67674	40961	46713	40065	51420
	$\overline{M}_w/\overline{M}_n$	3.21	3.07	3.07	4.25	4
	T_m (°C)	133	124	110	111	108
	T_c (°C)	116	110	91	93	102
	Crystallinity (%)	65	52	37	34	32

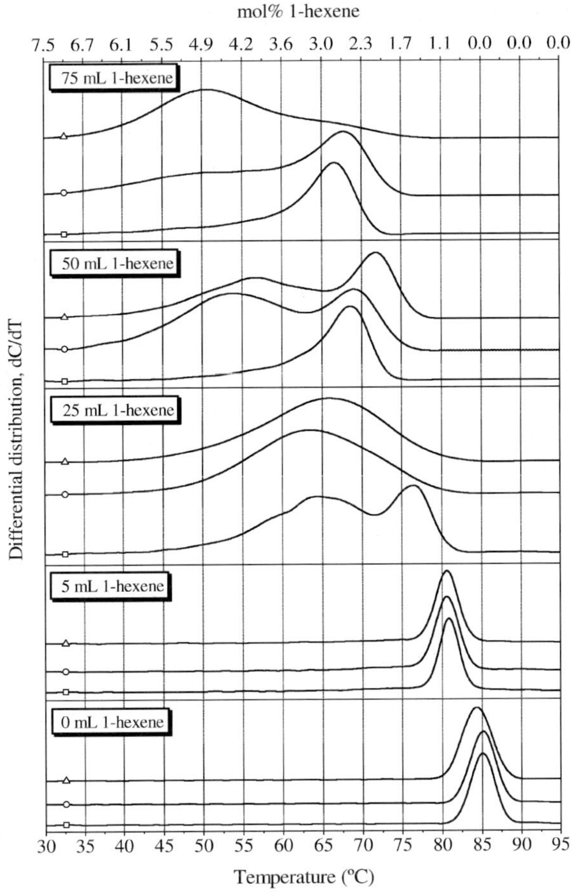

Figure 3.
Crystaf profiles of poly(ethylene-*co*-1-hexene) made with $(nBuCp)_2ZrCl_2$/MAO supported on MCM-41 with different Si/Al molar ratios: □ infinite; ○ 60 and △ 15. The mol% 1-hexene upper *x*-axis was calculated using the calibration curve in Figure 2.

Table 4.
Characterization of the poly(ethylene-*co*-1-hexene) made with $(nBuCp)_2ZrCl_2$/MAO immobilized on supports synthesized with different pore sizes.

Support	mL 1-hexene	0	5	25	50	75
SBA-15	$\overline{M}_n$ (g/mol)	74530	48869	55640	49333	68525
	$\overline{M}_w/\overline{M}_n$	2.67	2.47	2.67	3.01	3.22
	T_m (°C)	133	124	113	113	108
	T_c (°C)	116	111	95	93	88
	Crystallinity (%)	65	53	38	31	30
SBA-15-n-decane	$\overline{M}_n$ (g/mol)	53530	34968	43432	55751	58114
	$\overline{M}_w/\overline{M}_n$	2.78	2.66	2.69	3.03	3.44
	T_m (°C)	133	122	112	108	108
	T_c (°C)	117	109	94	89	91
	Crystallinity (%)	66	52	37	34	28
SBA-15-TMB	$\overline{M}_n$ (g/mol)	61391	44489	48885	59091	87092
	$\overline{M}_w/\overline{M}_n$	2.67	2.61	2.60	2.80	2.93
	T_m (°C)	133	123	110	106	102
	T_c (°C)	116	112	95	85	82
	Crystallinity (%)	65	50	37	29	26

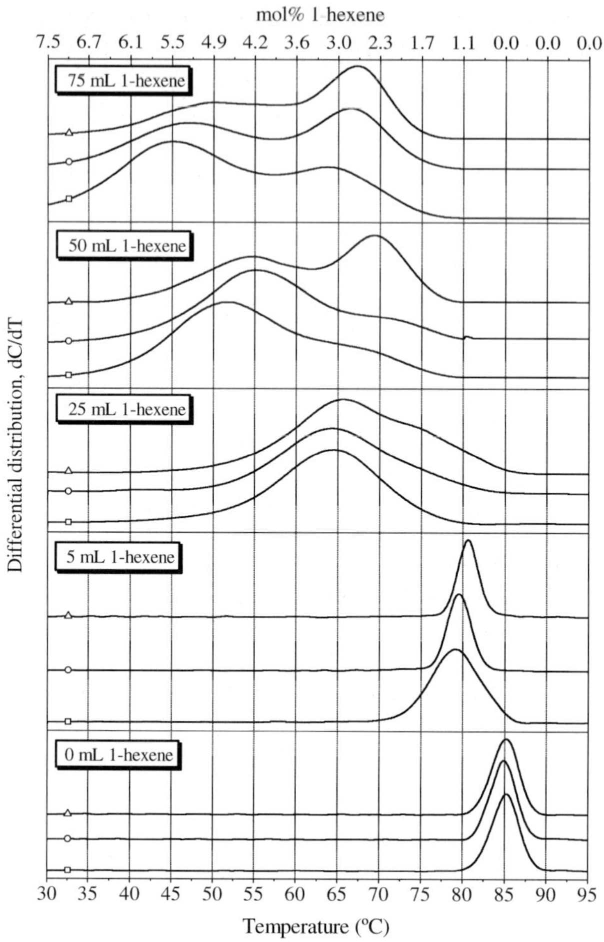

Figure 4.
Crystaf profiles of poly(ethylene-*co*-1-hexene) made with $(nBuCp)_2ZrCl_2$/MAO supported on SBA-15 with different pore sizes: □ 22.7 nm; ○ 11.3 nm and △ 8.8 nm. The mol% 1-hexene upper *x*-axis was calculated using the calibration curve in Figure 2.

Conclusions

All the poly(ethylene-*co*-1-hexene) samples made with $(nBuCp)_2ZrCl_2$/MAO supported on SiO_2, SiO_2/Al_2O_3, MCM-41 with different Si/Al ratios, or SBA-15 with different pore size, had narrow MWD, but CCDs that could vary from narrow and unimodal to broad and bimodal. The CCDs of copolymers made with $(nBuCp)_2ZrCl_2$/MAO supported on SiO_2 and SiO_2/Al_2O_3 were always unimodal, but those made with MCM-41 and SBA-15 became bimodal with increasing 1-hexene content. This bimodality may be related to the presence of two catalyst site types on the surface of the support. Lower Crystaf temperature peaks (that is, polymer populations with higher 1-hexene fractions) became more prominent with decreasing Si/Al ratio, and larger pore sizes seemed to favour 1-hexene incorporation slightly.

This investigation demonstrates that copolymer samples that do not differ significantly in their MWD shapes can have totally distinct CCD profiles as measured by Crystaf. Therefore, Crystaf (or equivalently, temperature rising elution fractionation,

Table 5.
Comparison of 1-hexene mol% of poly(ethylene-*co*-1-hexene) samples measured by ^{13}C NMR and calculated with the Crystaf calibration curve.

Support	mol% 1- hexene ^{13}C NMR	mol% 1- hexene calibration curve
MCM-41-15	0.79	1.05
	3.12	3.08
	3.54	3.77
	4.39	4.49
MCM-41-60	0.94	1.09
	3.10	3.17
	3.74	3.79
	3.75	3.71
MCM-41-infinite	0.80	1.04
	2.83	2.67
	2.90	2.91
	3.35	3.26
SBA-15	0.76	1.00
	2.67	2.68
	3.44	3.42
	3.72	3.72
SBA-15- n-decane	0.87	1.16
	2.87	2.95
	4.20	3.89
	4.22	4.11
SBA-15-TMB	0.79	1.21
	3.01	3.18
	4.32	4.24
	4.71	4.71

TREF) is an essential analysis to characterize copolymers made with coordination catalysts, even when they are expected to behave as single-site type catalysts.

[1] J. H. Z. dos Santos, A. Larentis, M. Barbosa da Rosa, C. Krug, I. J. R. Baumbol, J. Dupont, F. C. Stedile, M. D. C. Forte, *Macrom. Chem. Phys.* **1999**, 200(4), 751.

[2] G. Fink, R. Mülhaupt, H. H. Brintzinger, Ziegler Catalyst: Recent Innovations and Technological Improvements, Springer-Verlag, Berlin **1995**, 42.

[3] M. R. Ribeiro, A. Deffieux, M. F. Portela, *Ind. Eng. Chem. Res.* **1997**, 36(4), 1224.

[4] K.-J. Chu, J. B. P. Soares, A. Penlidis, *J. Polym. Sci., Part A: Polym. Chem.* **2000**, 38, 462.

[5] A. E. Hamielec, J. B. P. Soares, *Prog. Polym. Sci.* **1996**, 21, 651.

[6] G. G. Hlatky, *Chem. Rev.* **2000**, 100(4), 1347.

[7] Eur. Patent 336,593, M. Chang.

[8] U.S. Patent 4,912,075, M. Chang.

[9] K. Soga, M. Kaminaka, *Makromol. Chem. Rapid Commun.* **1992**, 13, 221.

[10] J. C. W. Chien, D. He, *J. Polym. Sci., Part A: Polym. Chem.* **1991**, 29, 1603.

[11] S. K. Ihm, K. J. Chu, J. H. Yim, In Catalyst Design for Tailor-Made Polyolefins, K. Soga, M. Terano, Eds., Elsevier-Kodansa, Tokyo **1994**, 299.

[12] M. Kaminaka, K. Soga, *Makromol. Chem. Rapid Commun.* **1991**, 12, 367.

[13] U.S. Patent 4,431,788, W. Kaminsky.

[14] Y. S. Ko, T. K. Han, J. W. Park, S. I. Woo, *Macromol. Rapid Commun.* **1996**, 17, 749.

[15] D. Y. Lee, K. B. Yoon, *Macromol. Rapid Commun.* **1994**, 15, 841.

[16] U.S. Patent 4,921,825, M. Kioka, N. Kashiwa.

[17] U.S. Patent 5,362,824, A. B. Furtek, R. S. Shinomoto.

[18] D. Trong On, D. Desplantier-Giscard, C. Danumah, S. Kaliaguine, *Appl. Catal. A-Gen.* **2001**, 222(1–2), 299.

[19] U. Ciesla, F. Schuth, *Micropor. Mesopor. Mat.* **1999**, 27, 131.

[20] W. Lin, Q. Cai, W. Pang, Y. Yue, B. Zou, *Micropor. Mesopor. Mat.* **1999**, 33, 187.

[21] D. Zhao, J. Feng, Q. Huo, N. Melosh, H. Fredrickson, B. Chmelka, G. D. Stucky, *Science* **1998**, 279, 548.

[22] S. Jüngling, S. Koltzenburg, R. Mülhaupt, *J. Polym. Sci., Part A: Polym. Chem.* **1997**, 35, 1.

[23] K. Soga, M. Kaminaka, *Makromol. Chem.* **1993**, 194, 1745.

[24] J. D. Kim, J. B. P. Soares, G. L. Rempel, *J. Polym. Sci., Part A: Polym. Chem.* **1999**, 37, 331.

[25] Y. Wang, M. Noguchi, Y. Takahashi, Y. Ohtsuka, *Catal. Today* **2001**, 68, 3.

[26] E. Byambajav, Y. Ohtsuka, *Appl. Catal. A-Gen* **2003**, 252(1), 193.

[27] Y.-H. Yue, A. Gedeon, J.-L. Bonardet, J. B. d'Espinose, N. Melosh, J. Fraissard, in: A. Sayari, M. Jaroniec, T. J. Pinnavaia, Eds., *Stud. Surf. Sci. Catal.* **2000**, 129, 209.

[28] D. Bianchini, K. M. Bichinho, J. H. Z. dos Santos, *Polymer* **2002**, 43(10), 2937.

[29] B. Monrabal, J. Blanco, J. Nieto, J. B. P. Soares, *J. Polym. Sci., Part A: Polym. Chem.* **1999**, 37, 89.

[30] D. M. Sarzotti, J. B. P. Soares, A. Penlidis, *J. Polym. Sci., Part B: Polym. Phys.* **2002**, 40, 2595.

[31] V. I. Costa, P. G. Belelli, J. H. Z. dos Santos, M. L. Ferreira, D. E. Damián, *J. Catal.* **2001**, 204, 1.

[32] Z. Ye, H. Alsyouri, S. Zhu, Y. S. Lin, *Polymer* **2003**, 44, 969.

DOI: 10.1002/masy.200751110

Application of Fractionation Techniques to the Study of Olefin Polymerization Kinetics and Polymer Degradation

Hisayuki Nakatani,[1] *Hitoshi Matsuoka,*[2] *Shoutarou Suzuki,*[3] *Toshiaki Taniike,*[3] *Liu Boping,*[3] *Minoru Terano*[*3]

Summary: Temperature rising elution fractionation (TREF) has been regarded as a powerful technique for study of semicrystalline polymers. In this paper, two examples of unique applications of TREF were introduced. One was the study on the influence of extraction of internal donor on the variation of isospecific active sites of a $MgCl_2$-supported Ziegler catalyst, and the other was the estimation of the relationship between polymer micro-tacticity and degradation rate of isotactic polypropylene (iPP). The former example revealed the conversion from high to low isospecific site by the extraction of internal donors, whereas the latter showed a negative correlation between the level of isotacticity and the degradation rate. These results demonstrated that TREF was useful in these research applications.

Keywords: catalyst; degradation; fractionation polymers; isotactic; polypropylene

Introduction

Analytical TREF has been a useful technique for the quantitative structural analysis of semicrystalline polymers.[1–3] In addition, TREF is a well-known procedure to fractionate each polymer component by differences of crystallizability. TREF has been regarded as the method of choice for the analysis and fractionation of polymer crystalline distribution.

It should be noticed here that the crystallizability of isotactic polypropylene (iPP) strongly depends on its isotacticity. In the case of iPP synthesized by $MgCl_2$-supported Ziegler catalysts, the iPP obtained often shows a broad distribution of isotacticities. This distribution directly reflects the existence of various isospecific active sites on the catalyst, with individual fractions of narrow isotactic distributions being produced by each corresponding active site type. TREF application for such iPP reveals the distribution profile of isospecific active sites on the Ziegler catalyst.[4,5]

Another substantial advantage of TREF is the fractionation of iPP with broad isotacticity distribution into several fractions of different isotacticities. In our previous studies,[6,7] the effect of tacticity on thermal oxidative degradation of polypropylene was investigated using highly isotactic, syndiotactic and atactic polypropylene resins, suggesting that the degree of oxidation of iPP was the severest. On the basis of this result, it was concluded that the tacticity affects one reaction path in the degradation process. However, the relationship between the level of tacticity and the rate of degradation had not been systematically clarified. The elucidation of this phenomenon requires the use of a systematic iPP series

[1] Kitami Institute of Technology, Koen-cho, Kitami, Hokkaido, 090-8507, Japan
[2] Tsukuba Research Laboratories, Tokuyama Corporation, 40, Wadai, Tsukuba-shi, Ibaraki, 300-4247, Japan
[3] Japan Advanced Institute of Science & Technology, Nomi, Ishikawa, 923-1292, Japan
Fax: (+51)761511625
E-mail: terano@jaist.ac.jp

with various isotacticities. By development of polymerization catalyst technology in recent years,[8–10] it is possible to control the tacticity of polypropylene. However, it is difficult to directly synthesize an iPP series with different isotacticities with only one kind of catalyst. The study of the relationship between degradation characteristics and isotacticity of iPP requires an iPP series with systematic isotacticity variation, and it has been difficult to synthetically supply such iPP series in practice. A better method is to fractionate a single iPP resin having broad isotacticity distribution witeh TREF to generate such a iPP series with various isotacticities.

In this paper, two kinds of TREF applications are reported: one to evaluate the influence of extraction of internal donor on the change of isospecificity distribution on $MgCl_2$-supported Ziegler catalysts and the other to clarify the relationship between the level of tacticity and the rate of degradation of iPP.

Experimental Part

Materials

Research grade propylene, donated by Tokuyama Corp., was used without further purification. $MgCl_2$ was kindly supplied by Toho Titanium Co., Ltd.

Propylene of research grade (donated by Tokuyama Corp and Chisso Corp.), anhydrous $MgCl_2$, $TiCl_4$, $TiCl_3$ (donated by Toho Catalyst Co.), nitrogen (purchased from Uno Sanso Co.), and triethylaluminum (TEA) (donated by Tosoh Finechem Co.) were used without further purification. Heptane, toluene, ethylbenzoate (EB) and dibutylphthalate (DBP) were purified by passage through a 13X molecular sieves column. TEA, EB and DBP were used as toluene solution.

Catalyst Preparation

Two types (Cat-A and Cat-B) of $TiCl_4/MgCl_2$ catalysts with two types of internal donors and one type of $TiCl_3/MgCl_2$ catalyst without any donors were prepared as follows.

Cat-A: $MgCl_2$ (36 g; 11 m^2/g), and EB (7.2 ml) were placed in a 1.2 L stainless steel vibration mill pot with 55 balls (25 mm diameter) under nitrogen and ground for 30h at room temperature. The ground product (200 ml) was treated with $TiCl_4$ (200 ml) in a 1.0 L three-necked flask at 90 °C for 2h with stirring under nitrogen, followed by washing several times with heptane and finally stored as a toluene slurry before polymerization. The Ti content of the catalyst was found to be 0.46 mmol-Ti/g-cat.

Cat-B: Cat-B was prepared by the same procedure as Cat-A, with DBP (7.2 ml) instead of EB. The Ti content of the catalyst was found to be 0.50 mmol-Ti/g-cat.

$TiCl_3/MgCl_2$: $TiCl_3/MgCl_2$ catalyst was prepared by grinding $MgCl_2$ and $TiCl_3$ in a 1.2 L stainless steel vibration mill pot with 55 balls (25 mm diameter) under nitrogen atmosphere for 30 h at room temperature. The Ti content of catalyst was 0.05 mmol-Ti/g-cat.

Extraction of Internal Donor

The $MgCl_2$ supported catalyst (about 1 g), TEA (14 mmol, Al/Ti mole ratio = 30) and toluene (200 ml) were placed in a 300 ml flask under nitrogen and the mixture was stirred at 30 °C for a selected time (from 1 to 30 min). The resulting catalyst was then recovered by filtration.

Propylene Polymerization and Estimation of Kinetic Parameters by Stopped-Flow Method

Stopped-flow polymerization of propylene and estimation of the kinetic parameters were carried out according to a previously reported method.[11–13] The propylene polymerization was typically performed with the catalyst (Cat-A or Cat-B: ca.1.0 g) and TEA (14 mmol, Al/Ti mole ratio = 30) in toluene at 30 °C for 0.15 s. The TEA solution in toluene (100 ml) saturated with propylene (1 atm) was placed into the vessel.

The propagation rate constant (k_p) and the concentration of active sites ([C*]) were

Table 1.
Yield, molecular characterization, and kinetic data of iPP.

Catalyst	Extract. time	Int. donor cont.[a)]	Yield	$M_n(M_w/M_n)$[b)]	mmmm[c)]	k_p	[C*]
	min	wt%	g/mol-Ti		mol%	l/mol · s	mol%
Cat-A	none	9.7	130	6100(3.3)	92.0	2700	2.15
	1	1.6	108	4800(3.8)	81.2	2400	2.01
	5	0.7	85	4700(4.0)	70.3	2100	1.81
	30	0.3	71	4000(4.2)	60.4	1600	1.75
Cat-B	none	12.8	53	6600(3.2)	94.0	3000	0.79
	1	8.7	45	5400(3.6)	88.4	2650	0.77
	5	8.0	42	5200(3.8)	85.2	2500	0.75
	30	5.5	37	4700(4.2)	78.5	2300	0.72

Polym. condition: Al/Ti = 30, polym. temp. = 30 °C, polym. time = 0.15s;
a) Amounts of remaining internal donor on catalyst after extraction experiment were determined by GC;
b) Molecular weight and its distribution were determined by GPC;
c) Meso pentad fraction was determined by ^{13}C-NMR.

determined by the following equations:

$$M_n = M_0 * \frac{k_p * [M] * t}{1 + k_{tr} * t} \quad (1)$$

$$Y = k_p * [M] * [C*] * t \quad (2)$$

where M_n, M_0, [M], t, k_{tr}, and Y are number-average molecular weight of polymer, molecular weight of monomer, monomer concentration, polymerization time, transfer rate constant, and polymer yield, respectively.

The iPP having broad isotacticity distribution was synthesized using $TiCl_3/MgCl_2$ catalyst and TEA. Polymerization was performed at an Al/Ti molar ratio of ca. 2 at 30 °C for 30 min in toluene. In order to remove catalyst residues, the obtained iPP was reprecipiated from a boiling xylene solution into methanol under nitrogen atmosphere.

Temperature Rising Elution Fractionation (TREF)

The isotacticity distribution of iPP polymerized with the $TiCl_3/MgCl_2$ catalyst system was determined by TREF (Senshu SSC-7300) using *o*-dichlorobenzene (ODCB) containing antioxidant (*2,6-di-t*-butyl-*p*-cresol) as the extraction solvent. A fractionation column packed with Chromosorb (Celite Corp.) with 10 mm diameter and 30 cm in length was used for the TREF characterization. About 1.4 g of iPP was dissolved in 70 ml of ODCB at 140 °C. A

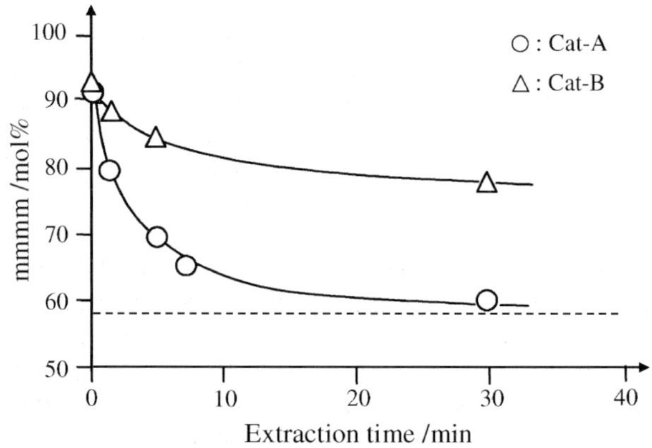

Figure 1.
Plots of extraction time versus isotacticity of iPPs obtained from Cat-A and Cat-B: Dashed line represents isotacticity of iPP obtained from no internal donor catalyst.

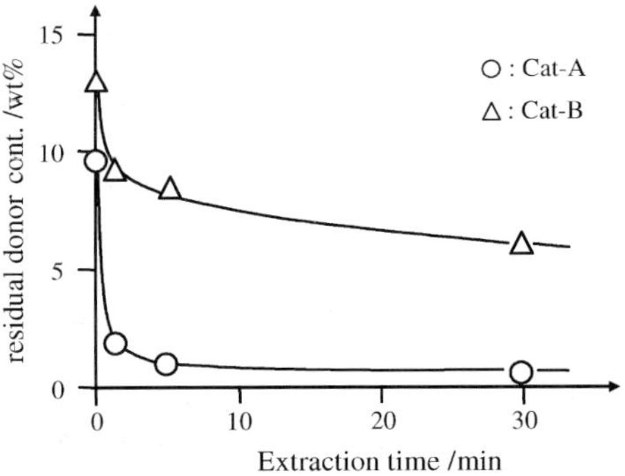

Figure 2.
Content of residual internal donor as a function of the extraction time.

part of the solution (50 ml, ca. 20 mg/ml) was eluted through the fractionation column. The column was then cooled to 20 °C slowly (6.7 °C/h). Elution with ODCB (150 ml/h) was first carried out at 20 °C for 30 min to obtain the ODCB-soluble fraction, and then the column was heated up to 140 °C at 10 °C/h.

The precise fractionation of iPP was carried out with the same TREF device, solvent, and column. About 70 mg of the iPP was dissolved in 10 ml of ODCB at 140 °C, and a part of the solution (ca. 6 ml) was passed through the fractionation column, which was slowly cooled down at 6.7 °C/h from 140 °C to 20 °C. Elution of the deposited iPP with ODCB at a flow rate of 150 ml/h was first carried out at 20 °C for 30 min to obtain the ODCB-soluble fraction, and then the column was heated at 16 °C/h up to 140 °C. The eluted iPP solution was analyzed by a refractive index detector to obtain the TREF diagram. Moreover, tacticity fractions of the iPP with increasing crystallinity were then eluted with ODCB at the temperatures increasing stepwise from 20 °C to 140 °C divided into 5 steps (20, 60, 90, 106 and 140 °C). Polymer was eluted during 90 min at every step after the temperature had stabilized for 30 min. The fraction was then collected and precipitated in methanol. All fractionated iPP were TREF fractions insoluble below 20 °C. The weight fraction of the soluble part (below 20 °C) was 52 wt%, and its molecular weight and tacticity were very low ($M_n = 6{,}000$ and mmmm = 25 mol%). Since this soluble part was considered to contain impurities, such as unsaturated polymer chain end groups, for promoting degradation,[10] it was completely removed.

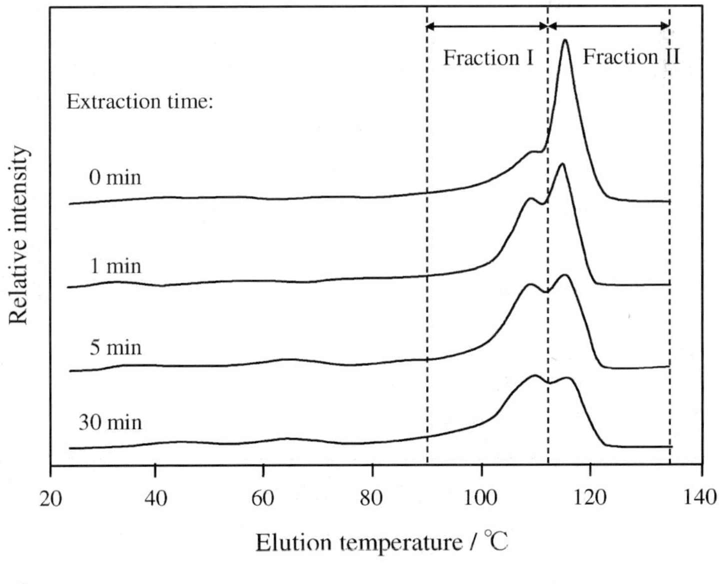

Figure 3.
TREF diagrams of iPPs obtained from the extracted Cat-A.

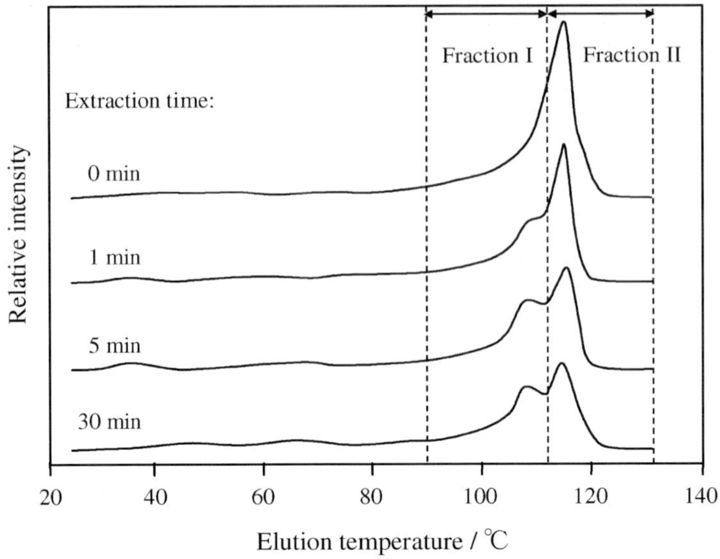

Figure 4.
TREF diagrams of iPPs obtained from the extracted Cat-B.

GPC Characterization

The sample in the small vial was dissolved in 5ml of ODCB containing 2,6-di-t-butyl-p-cresol as an antioxidant, and the obtained solution was directly measured by GPC. The molecular weight was determined by GPC (Senshu, SSC-7100) with styrene-divinylbenzene gel columns (SHODEX, HT-806M) at 140 °C using ODCB as a solvent.

^{13}C-NMR Measurement

Isotacticity was determined by ^{13}C-NMR measurement using a Varian Gemini-300 spectrometer at 120 °C on 20 % (w/v) solution in 1,2,4-trichlorobenzene or hexachloro-1,3-butadiene. Benzene-d_6 or 1,1,2,2-tetrachloroethane-d_2 was added as an internal lock and was used as an internal chemical shift reference.

Thermal Oxidative Degradation

Thermal oxidative degradation of the fractionated iPP was performed at 130 °C for 10–120 min in air using a Senshu SSC-9300 heater. No additional antioxidant was used in any cases. The iPP sample was first dried

Table 2.
Characterization and kinetic data of Fraction I and Fraction II in the iPP obtained from extracted Cat-A catalyst*.

Fraction	Extract. time	M_n	mmmm	k_p	[C*]
	min		mol%	l/mol · s	mol%
	0	10 500	95.0	4700	10
	1	10 500	94.7	4700	13
Fraction I	5	11 000	94.3	4900	18
	30	10 500	94.5	4700	24
	0	27 800	98.6	12 500	40
	1	27 600	98.2	12 300	36
Fraction II	5	28 000	98.4	12 500	30
	30	27 400	98.5	12 200	16

b) *Polymerization condition and analytical methods are the same as Table 1.

Table 3.
Characterization and kinetic data of Fraction I and Fraction II in the iPP obtained from extracted Cat-B catalyst*.

Fraction	Extract. time	M_n	mmmm	k_p	[C*]
	min		mol%	l/mol · s	mol%
	0	10 600	95.2	4800	8
	1	10 700	95.1	4900	9
Fraction I	5	10 500	95.4	4700	11
	30	10 500	94.8	4700	14
	0	27 500	98.5	12 300	22
	1	27 400	98.5	12 200	19
Fraction II	5	27 300	98.6	12 200	15
	30	27 600	98.7	12 300	10

a *Polymerization condition and analytical methods are the same as Table 1.

in vacuo at 60 °C for 3 h. Then, 5 mg of each iPP sample placed in a small vial was transfer to the heater and covered with a heating jacket to achieve effective heat transfer to the iPP sample.

Thermogravimetric Analysis

Decrease in weight of iPP was analyzed by thermogravimetric analysis (TGA, METTER TG50) at 130 °C for 40 h under air (air flow rate: 50ml/min.).

Results and Discussion

TREF Application for Polymerization Kinetics

Polypropylene was prepared by the stopped-flow method. It is noticed here that this method applies very short polymerization times in which chain transfer reactions can be disregarded to simplify Equation (1).[11] Characterization and kinetic data of the obtained iPPs were summarized in Table 1. All yields and M_n decrease with the treatment time by TEA, suggesting that the extraction of internal donor critically affects the concentration of active sites ([C*]) and k_p in $TiCl_4/MgCl_2$ catalyst systems.

Figure 1 plots extraction time versus isotacticity. The isotacticity decreases with increasing extraction time. It seems that Cat-A and Cat–B loose their isospecific active sites in the extraction process. In the case of the $TiCl_4/MgCl_2$ catalyst without internal donor, the isotacticity (mmmm) was almost 58 mol% under the same polymerization conditions (see the dashed line in Figure 1). As shown in the figure, the isospecificity of Cat-A rapidly reaches that of the catalyst without internal donor, indicating that the interaction between active sites and EB is weaker than that between active sites and DBP.

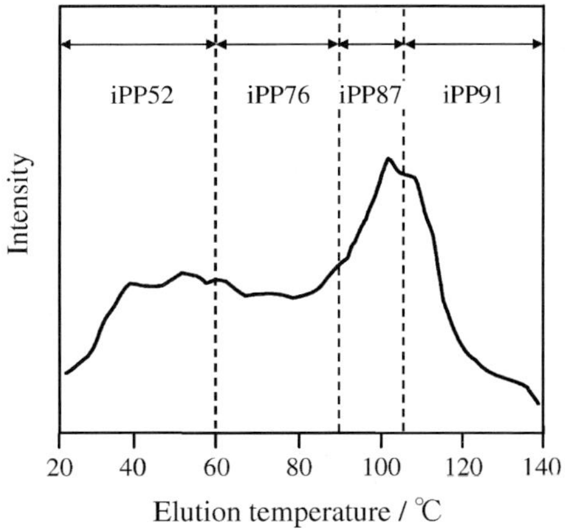

Figure 5.
TREF diagram of iPP with broad isotacticity distribution.

The content of the residual internal donor as a function of extraction time is shown in Figure 2. Here the contents of these residual donors were determined by gas chromatography. EB was extracted rapidly from the catalysts. The extraction rate of EB is higher than that of DBP, demonstrating that the interaction between EB and $MgCl_2$ is weaker. It is noticed here that this rapid change is apparently different from the rather slow one for isospecificity. Although donor extraction certainly causes a decrease in the number of isospecific active sites, it can not account

Table 4.
Fraction ratio and molecular weight of fractionated iPPs.*

Sample	Range of elut. temp. /°C	Fraction ratio /wt%	M_n	M_w/M_n
—	20	52	0.6 × 104	—
iPP52	20–60	17	1.1 × 104	4.6
iPP76	60–90	12	1.8 × 104	5.1
iPP87	90–106	9	2.5 × 104	3.2
iPP91	106–140	10	7.4 × 104	8.5

b *Analytical method is the same as Table 1.

Table 5.
^{13}C-NMR pentad distribution (mol%) for iPPs.

Samples	mmmm	mmmr	rmmr	mmrr	mmrm+rrmr	mrmr	rrrr	mrrr	mrrm
iPP52	51.5	13.3	1.7	12.2	5.6	1.4	5.2	4.1	5.0
iPP76	76.3	9.8	0.0	7.0	2.0	0.0	1.8	0.6	2.5
iPP87	87.0	6.2	0.7	3.2	0.7	0.1	0.6	0.4	1.1
iPP91	90.6	6.5	0.4	1.4	0.3	0.1	0.1	0.1	0.5

entirely for such slow decrease in isospecificity.

The application of TREF visually elucidates the changes in the distribution of isospecific active sites shown in Figure 3 and 4. There exist two remarkable peaks in these TREF diagrams. The higher temperature (ca. 115 °C) peak corresponds to a higher isospecific active site, and the lower (ca. 108 °C) to lower isospecific site. The intensity of the higher peak gradually decreases with increasing extraction time. Interestingly, the lower peak grows with increasing extraction time of the internal donor. The growing suggests that the higher isospecific active site is converted to the lower one by the extraction of the internal donor. In order to study this phenomenon in detail, the iPPs were fractionated into two fractions (denoted as Fraction I and Fraction II) in two temperature ranges of 90 ~ 112 °C and 112 ~ 135 °C, respectively. Fractions I and II are produced with the lower and higher isospecific active site, respectively, permiting the estimation of the [C*] for each site type. The characterization and the kinetic data are summarized in Table 2 and 3. It appears that the [C*] of the lower isospecific site increases as the [C*] for the higher isospecific site decreases. In particular, this tendency is remarkable in the iPP series made with Cat-A, in good agreement with the changes in the TREF profiles. The TREF results demonstrate that the higher isospecific active site is converted into the lower one by the extraction of the internal donor.

In conclusion, TREF has been proved to be quite effective to evaluate the change of stereospecificity of active sites.

TREF Application to Polymer Degradation Behavior

PP is very vulnerable against oxidation. This process is called "degradation" and it is a very important practical issue. Although many researchers have been involved in developing industrial polypropylene resins,

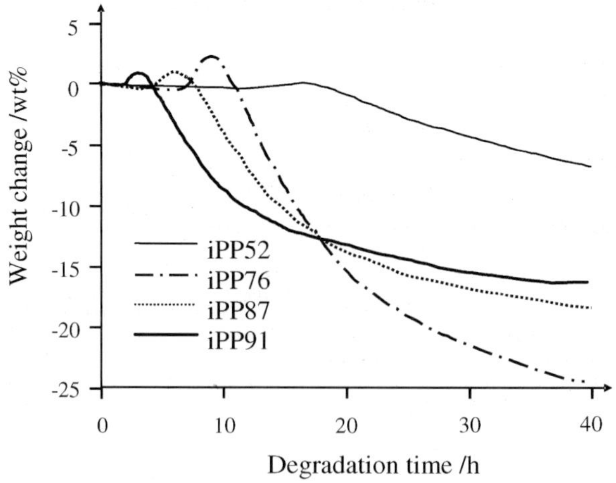

Figure 6.
Thermogravimetric analysis results of fractionated iPPs.

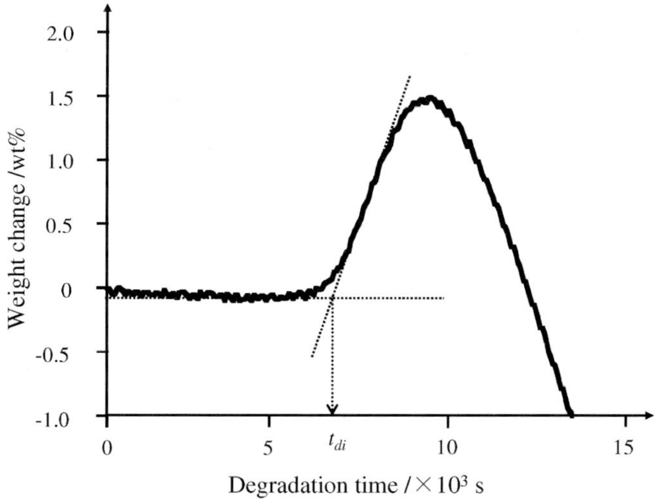

Figure 7.
An example of weight change curve showing a determination of the durations of induction period (t_{di}).

polypropylene degradation has not been comprehensively studied with the viewpoint of a stereochemical reaction. The reason for this is that it is difficult to prepare a series of polypropylene resins with various tacticities.

As shown in Figure 5, we succeeded in the synthesis of iPP with broad isotacticity distribution using the $TiCl_3/MgCl_2$ catalyst.[9] Four kinds of iPP with different isotacticities were easily obtained from the fractionation of this iPP by TREF. The fraction ratios and molecular weights are summarized in Table 4. The nomenclature of the fractionated iPP is as follows: iPP91 denotes iPP with the pentad fraction (mmmm) of ca. 91 mol%. All pentad fractions of the iPPs are shown in Table 5.

The dependence of thermal oxidative degradation on the isotacticity was carried out with these iPPs. It is noticed here that all of the thermal oxidations have been

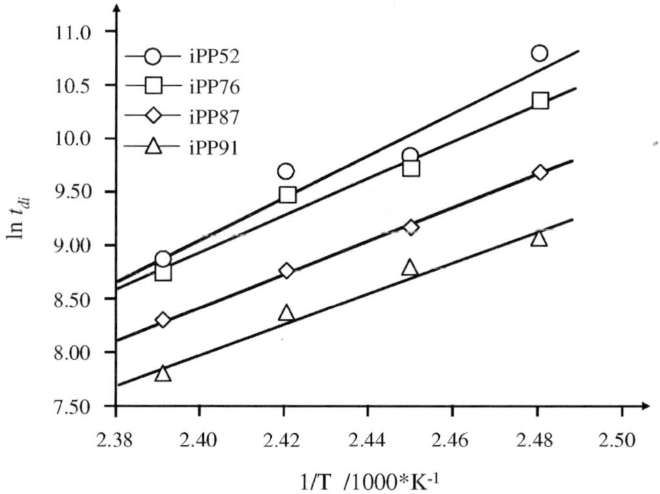

Figure 8.
Arrhenius plots of t_{di} for thermal oxidative degradation of iPPs.

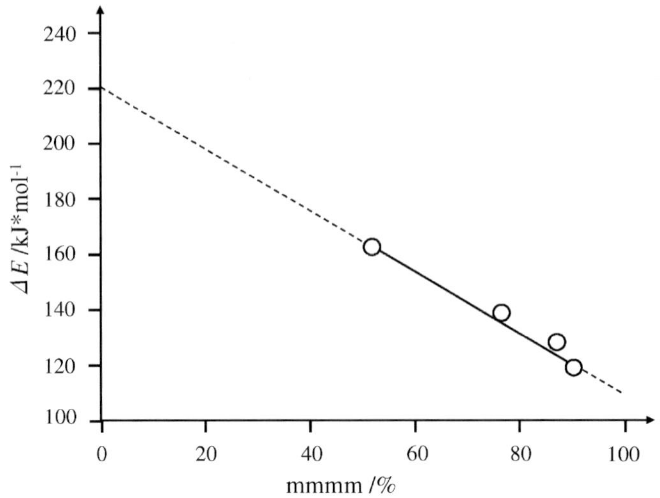

Figure 9.
Activation energy (ΔE) versus meso pentad fraction (mmmm) for all of degraded iPPs.

performed at 130 °C. At such higher temperature,[14,15] the degradation initiator (radical species) can be diffused sufficiently in the iPP crystalline part, and the degradation reaction proceeds homogeneously in the whole of the iPP sample.

These weight change curves measured by TGA are shown in Figure 6. The thermal oxidative degradation was performed at a constant temperature (130 °C) in atmospheric air. The weight changes of iPPs display multi-stages with advance of degradation.[16,17] A period of constant weight, namely induction period, is initially seen, and then the weight is increasing by oxidation. Beyond a given maximum wight, weight loss starts to take place as some components are volatilized, and finally the mass balance becomes negative. Figure 6 shows that the weight increase, or end of the induction period, starts earlier as the mmmm becomes higher. These results show that the initiation of the degradation reaction depends strongly on the isotacticity. As illustrated in Figure 7, the duration of induction period (t_{di}) was defined as the time (in seconds) at the intersection of the tangent at the inflection point with the plateau line of induction period. The t_{di} was determined using the weight change curves obtained from the measurements at four constant temperatures (130, 135, 140 and 145 °C). These values have been fitted to an Arrhenius expression to obtain the apparent activation energy (ΔE). As shown in Figure 8, the t_{di} apparently obeys the Arrhenius law within the chosen temperature limits. All the obtained ΔE are plotted against the mmmm in Figure 9.

It appears that ΔE is inversely proportional to mmmm. The relationship obeys the following equation:

$$\Delta E(\mathrm{kJ/mol}) = -1.24 \times [\mathrm{mmmm}] + 220 \tag{3}$$

This result implies that the stability of iPP can be predicted without thermal degradation testing using the simple Equation (3). Actually the availability of this equation has been confirmed by using a higher iPP (mmmm = 95 mol%, $\overline{M}_n = 1.8 \times 10^4$ g/mol, $\overline{M}_w/\overline{M}_n = 4.9$).

Therefore, we demonstrated that TREF can be very useful for the study the influence of isotacticity on the degradation of polypropylene.

Conclusions

TREF has been regarded as a powerful technique for the structural study of semicrystalline polymers. In this paper, two examples of TREF applications were reported for isospecific propylene polymerization kinetics and for degradation behavior of isotactic polypropylene. TREF could detect the conversion from higher to lower isospecific sites by extraction of internal donor and estimate an inverse correlation between isotacticity and degradation rate of isotactic polypropylene. It was found that TREF possesses an extensive applicability to different problems in polymer science.

[1] M. Kakugo, T. Miyatake, K. Mizunuma, Y. Kawai, *Macromolecules* **1988**, *21*, 2309.
[2] M. Kakugo, T. Miyatake, T. Mizunuma, *Macromolecules* **1991**, *24*, 1469.
[3] M. Kioka, H. Makio, A. Mizuno, N. Kashiwa, *Polymer* **1994**, *35*, 580.
[4] H. Matsuoka, B. Liu, H. Nakatani, M. Terano, *Macromol. Rapid Commun.* **2001**, *22*, 326.
[5] H. Matsuoka, B. Liu, H. Nakatani, I. Nishiyama, M. Terano, *Polym. Int.* **2002**, *51*, 781.
[6] S. Suzuki, Y. Nakamura, A. K. Hasan, B. Liu, M. Terano, H. Nakatani, *Polym. Bull.* **2005**, 54(4–5), 311.
[7] H. Nakatani, S. Suzuki, T. Tanaka, M. Terano, *Polymer* **2005**, *46*(26), 12366.
[8] H. H. Brintzinger, D. Fischer, R. Mülhaupt, B. Rieger, R. M. Waymouth, *Angew. Chem. Int. Ed. Engl.* **1995**, *34*, 1143.
[9] F. Grisi, P. Longo, A. Zambelli, J. A. Ewen, *J. Mol. Catal.* **1996**, *140*, 225.
[10] M. Matsuoka, S. Matsui, M. Mitani, J. Saito, K. Tsuru, N. Kashiwa, T. Fujita, *J. Mol. Catal.* **2001**, *169*, 99.
[11] T. Keii, M. Terano, K. Kimura, K. Ishii, *Makromol. Chem.,Rapid Commun.* **1987**, *8*, 583.
[12] H. Mori, H. Saito, M. Terano, *Macromol. Chem. Phys.* **1988**, *199*, 55.
[13] H. Matsuoka, B. Liu, H. Nakatani, M. Terano, *Macromol. Rapid Commun.* **2001**, *22*, 326.
[14] L. Achimsky, L. Audouin, J. Verdu, J. Rychly, L. Matisova-Rychla, *Polym. Degrad. Stab.* **1997**, *58*, 283.
[15] P. Tiemblo, J. M. Gómez-Elvira, S. Garcia Beltrán, L. Matisova-Rychla, J. Rychly, *Macromolecules* **2002**, *35*, 5922.
[16] L. Audouin, L. Achimsky, J. Verdu, in: "*Handbook of Polymer Degradation*", 2d ed., S. Halim Hamid, Ed., Marcel Dekker, New York 2000, p. 727.
[17] J. Rychly, L. Matisova-Rychla, K. Csmorova, L. Achimsky, L. Audouin, A. Tcharkhtchi, J. Verdu, *Polym. Degrad. Stab.* **1997**, *58*, 269.

DOI: 10.1002/masy.200751111

Synthesis and Characterization of Ethylene/Propylene Copolymers in the Whole Composition Range

Mª. Joaquina Caballero,[1] Inmaculada Suarez,[1] Baudilio Coto,[1] Rafael Van Grieken,[1] Benjamín Monrabal[2]*

Summary: The incorporation of comonomer molecules in the backbone of a homopolymer can influence the final properties of the material, decreasing its crystallinity and the melting and glass transition temperatures, and increasing its impact resistance and transparency. In the present work, ten ethylene/propylene copolymers have been synthesized using a supported metallocene catalytic system covering the whole composition range. Any desired composition was obtained by controlling the feed composition during the reaction. These synthesized copolymers have been characterized by different techniques in order to study the effect of the comonomer incorporation onto their final properties. When the comonomer content is low, the behaviour of the copolymer is similar to that of the corresponding homopolymer. Nevertheless, if the comonomer content increases, the copolymer becomes more amorphous (low crystallization temperature and soft XRD signals) and easily deformable, reaching a behaviour close to that corresponding to an elastomeric material. In order to corroborate these results the samples have been characterized by TREF and GPC-MALS. TREF analysis showed that copolymers containing less than 10% and more than 80% of ethylene are semicrystalline, with elution temperatures typical of this kind of polymers. Molecular weights are higher for homopolymers and they decrease as the comonomer concentration increases, whereas the polydispersity index keeps almost constant at the expected value for this kind of samples.

Keywords: ethylene/propylene copolymers; gel permeation chromatography (GPC); metallocene supported catalysts; multiangle light scattering (MALS); temperature rising elution fractionation (TREF)

Introduction

The thermoplastic polymer production with elastomeric properties has been the focus of scientists and manufacturers in the last decade.[1] Polypropylene (PP) is a semicrystalline polymer widely used in packaging, textile, and automobile industries because of its good processability and properties. Nevertheless, its applications are limited because PP toughness is low, especially at room and low temperatures.[2] This is the reason why the copolymerization products of propylene with small amounts of ethylene are of practical importance since they allow to control the crystallization processes.[3]

The development of metallocene catalyst technology has provided a rich set of new copolymers materials with well-defined microstructures.

In order to synthesize these products, the combination of single site catalysts with methylaluminoxane (MAO) as cocatalyst for olefin polymerization throughout the last two decades has been the subject of intensive research, as a result of high

[1] Department of Chemical and Environmental Technology, ESCET. Universidad Rey Juan Carlos, Móstoles, Madrid, Spain
[2] Polymer Char, Valencia Parc Tecnologic, Paterna, Valencia, Spain
E-mail: baudilio.coto@urjc.es

catalytic activity and versatile control of polymer structure and properties via manipulation of the ligand sphere surrounding the metal atom[4] allowing the preparation of homogeneous products covering a wide range of copolymer microstructures and properties.[5]

It is of great importance[6] to immobilize the homogeneous metallocene catalysts on inorganic carriers[4] for applying them to gas or slurry phase[7] polymerization processes because supported catalysts can avoid reactor fouling, allow the reduction of the amount of cocatalyst during the reaction,[8] and control the morphology of the produced polymer more easily.[9] The anchorage of the catalytic system can be carried out by several well-known techniques.[10,11]

In addition to the catalytic system used, the properties of ethylene-propylene copolymers also depend on comonomer content (composition) and its distribution, which, in turn, essentially depend on polymerization conditions.[12] Such conditions have to be kept constant over the entire polymerization period,[3] with a good control of the feed monomer ratio to obtain a copolymer with a defined composition in a semi-batch reactor.

The influence of these properties can be studied by analyzing the samples through different techniques. The use of gel permeation chromatography (GPC) with light scattering detector (MALS) provides quantitative information on molecular weight distribution and conformation of the chains; this technique has been studied intensively in recent years,[13–15] but there are not publications for ethylene/propylene copolymers. The behaviour of semicrystalline copolymers can be studied using different techniques. Differential scanning calorimetry (DSC) allows the measurement of transition temperatures and crystallinity percentage. Such crystallinity is related with the signals obtained by X-ray diffraction (XRD). Temperature rising elution fractionation (TREF) yields the qualitative estimation of chemical composition distribution (CCD) which is also related to crystallinity.

In this work, the catalytic system MAO/rac-$Me_2Si[2Me\text{-}Ind]_2ZrCl_2$ was impregnated on a thermally-treated silica support. This catalyst was used in ethylene-propylene copolymerization reactions. The polymerization processes was carried out with a variable and controlled feed of the monomers which allowed the synthesis process to take place in a wide range of relative concentration of the different monomers in the solution.

In order to check the effect of the copolymers composition on their properties, the copolymers were analyzed by X-ray diffraction (XRD), differential scanning calorimetry (DSC), ^{13}C nuclear magnetic resonance (^{13}C-NMR), scanning electron microscopy (SEM), temperature rising elution fractionation (TREF), and gel permeation chromatography coupled to multiangle light scattering (GPC-MALS).

The results show how the behaviour of the copolymers is typical of an elastomeric material when the percentage of both monomers is similar, while they are very close to the homopolymer behaviour when the content of one of the monomers is much higher than the other.

Experimental Part

Synthesis

The chemicals used were ethylene (Air Liquide S.A., polymerization grade 99.99%), propylene (Air Liquide S.A., polymerization grade 99.99%) and n-heptane (Scharlab S. A., 99%) as solvent. The metallocene cata lyst used was rac-dimethyl-silylbis (2-methylindenyl) zirconium dichloride (Boulder Scientific Company) supported on silica (Grace Davidson) modified with methylaluminoxane, MAO (Witco, 30 wt% in toluene) and the cocatalyst added to the reactor was triisobuthylaluminium, TIBA (Witco, 1 M in toluene).

All the air-sensitive compounds were handled in a dry nitrogen atmosphere inside a glove box and by using Schlenk techniques.

The conditions of the support pre-treatment, the determination of the hydroxyl content and the physical properties of the silica in these conditions were reported previously.[16]

The immobilization method was carried out in two steps: 1) reaction between 1 g of silica material and MAO (30 wt % in toluene) solution to a volume ratio of three, at room temperature and 30 rpm in a stirred vessel; 2) anchorage of the metallocene on 2 g of the silica/MAO sample previously synthetized. The catalyst (rac-$Me_2Si[2MeInd]_2ZrCl_2$) was dissolved in 75 mL of toluene and this mixture was kept at room temperature under stirring (900 rpm) for three hours.

Copolymerization reactions were carried out in a 1 liter Büchi stirred glass reactor which allows the control of the temperature during the reactions at 70 °C. The solvent (400 cm^3 of n-heptane) was saturated with an ethylene/propylene gas mixture previously to the copolymerization reaction. The Al/Zr molar ratio used in all the reactions was 400.

Ethylene and propylene were deoxygenated and dried through columns containing R-3/15 BASF catalyst, alumina and 3 Å molecular sieves before entering the polymerization reactor. The monomers were fed in order to keep the reactor pressure at 5 bar during the entire copolymerization.

The ethylene/propylene ratio in the gas phase (C_2/C_3) was kept constant. Real-time monitoring of the gas phase composition was performed by means of a Micro-GC (MGC supplied by Varian). The reactions were carried out during 30 minutes, and finally, the copolymers were precipitated by acidic methanol, washed and dried in vacuum.

Characterization

Copolymers synthetized were observed by SEM (XL30 ESEM) in order to study the replication phenomenon and the external morphology. Crystallinity was analyzed by powder X-ray diffraction (Philips X'PERT MPD diffractometer using Cu K_α). Melting points and glass transition temperatures were measured by differential scanning calorimetry (DSC, 822^e METTLER TOLEDO).

^{13}CNMR spectra were recorded in deuterated 1,1,2,2-tetrachloroethane and 1,2,4-trichlorobenzene at 100 °C using a BRUKER AC300 spectrometer and the final composition of the copolymers was calculated by ^{13}C-NMR triad distribution using known methods.[17,18]

GPC-MALS is a combined method of gel permeation chromatography (Waters ALLIANCE GPCV 2000) with multi-angle light scattering (DAWN EOS 18 angle light scattering photometer, Wyatt Technology). Polymer conformation can be obtained by GPC-MALS by direct application of the Zimm-Stockmayer approach.[19] Solvent was 1,2,4-trichlorobenzene (TCB) at a flow rate 1 mL/min and the temperature was 145 °C. At each chromatographic slice, both the absolute molecular weight (M_w) and radius of gyration (R_g) can be obtained.[20] The refractive index increment d*n*/d*c* value for these copolymers in TCB and 145 °C is 0.101 mL/g.[21]

Chemical composition distributions were measured by TREF 200 (Polymer Char) using o-dichlorobenzene (oDCB) as a solvent. In the elution step oDCB flowed through the column at a constant flow rate of 0.5 mL/min with a temperature profile from 35 °C to 140 °C at a constant rate (1 °C/min). The concentration of polymer was measured by a two-channel infrared detector.

Results and Discussion

Table 1 lists the copolymers obtained. They were named EP followed by an integer number related to the molar percentage value of ethylene. As can be seen, in the synthesis procedure the gas phase composition C_2/C_3 was controlled in the range 0.010 to up to 100 and that lead to copolymers in the whole composition range.

The external morphology of the samples was observed by SEM. Three SEM micrographs for different copolymers are shown in Figure 1. When supported catalytic

Table 1.
Obtained copolymers.

Copolymer	C_2/C_3 (mol/mol) gas phase composition	% E molar[a)]
EP3	0.010	3.4
EP6	0.013	5.8
EP13	0.029	12.7
EP21	0.126	20.5
EP31	0.390	30.4
EP59	0.970	59.0
EP81	5.046	80.7
EP87	6.740	86.7
EP94	19.000	93.8
EP99	107.69	99.0

a) Copolymer composition obtained from triad distribution as E = EEE+EEP+PEP.

systems are used, it is expected that the polymer particles replicates the shape of the support.[22] This behaviour can be seen in Figure 1.a and 1.b, with independent particles distribution, for copolymers where one of the monomers is the main component in the backbone of the chain. However, when the percentage of both monomers in the copolymer chain is similar, the final product does not present a defined morphology, as can be seen in the SEM image corresponding to the EP59 sample (Figure 1.c). This characteristic is typical of polymers synthesized with a dissolved catalytic system.[23] Such different behaviour could be related to the amorphous character of the material: EP59 is elastomeric and the support particle can not be broken when the polymerization takes place.

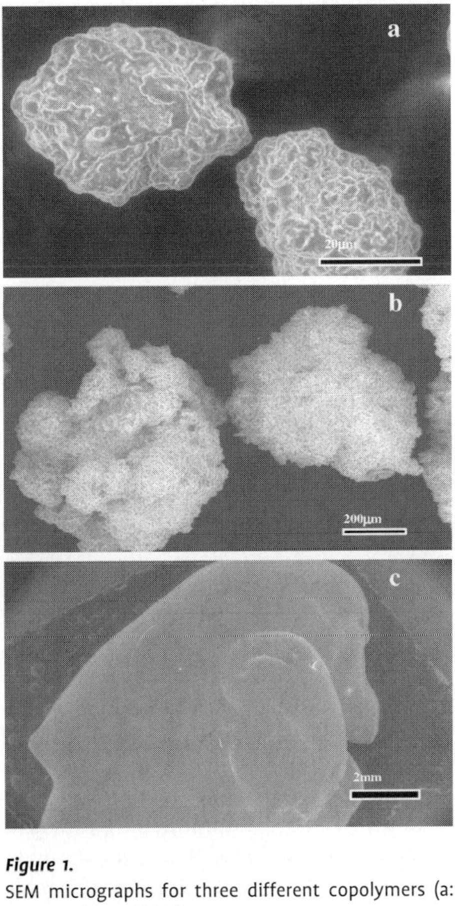

Figure 1.
SEM micrographs for three different copolymers (a: EP99, b: EP3, c: EP59).

The crystallinity of the synthesized samples has been studied qualitatively by XRD. The results are shown in Figure 2 for all the copolymers. It is possible to check how for copolymers EP99 and EP3 the typical peaks for polyethylene and polypropylene, respectively, are obtained. When the propylene is the monomer with higher percentage (Figure 2, EP3), it is possible to distinguish four fundamental signals in the spectra for 2θ values of 14.1, 16.9, 18.6, and 21.7, associated to the α or monoclinic modification of iPP.[24]

The peaks are wider when the ethylene content increases because the second monomer inclusion introduces higher disorder into the lattice. The maximum disorder is found for the EP59 copolymer, where it is impossible to distinguish any characteristic peak of the homopolymers.[12] On the other composition limit, when the ethylene content is larger, two signals can be observed which could be related with the orthorhombic[25] phase for 2θ values of 21.5° and 23.9°, respectively.

Figure 3 shows the results obtained by DSC for the glass transition temperature (T_g) and the melting temperature (T_{m2}) in function of the ethylene content. It is clear that values for these temperatures are lower for copolymers close to 50/50 ethylene/propylene ratio. This is related to the considerable decrease of crystallinity, since the crystal domains which are being formed

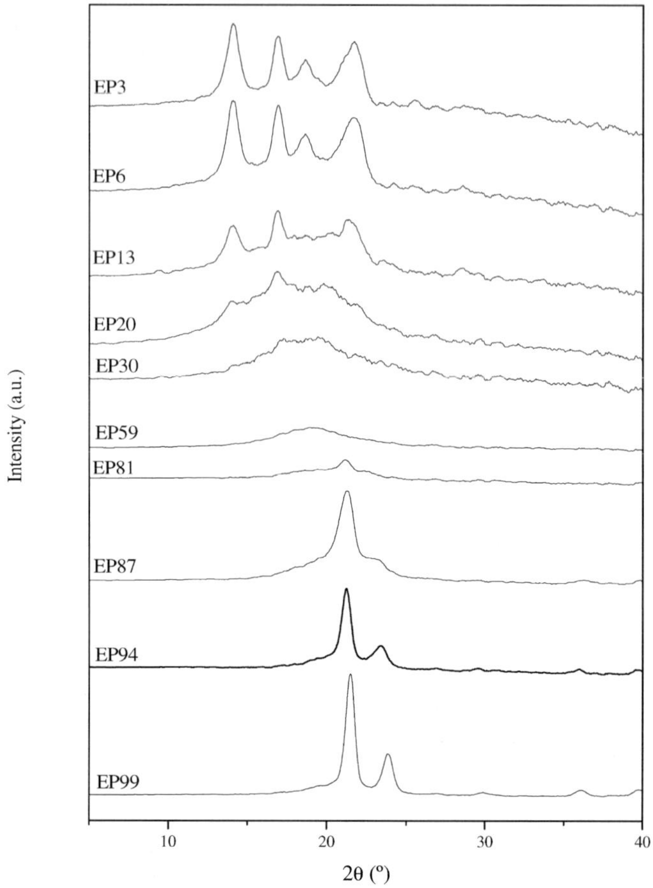

Figure 2.
Qualitative comparison of the crystallinity determined by XRD for copolymers with different percentages of ethylene.

are smaller because the increase of irregularities in the macromolecular chains.

From DSC results, values for the enthalpy of this second melting of the copolymers are obtained. It is possible to determine the percentages of crystallinity by relating such values to those for the corresponding homopolymers (291.6 J/g for polyethylene[26] and 207.1 J/g for polypropylene[27]). The calculated values are listed in Table 2 and corroborate how the copolymers with close to 50/50 ethylene/propylene ratio are amorphous materials, in agreement with the characterization by XRD.

From the experimental triad values for the copolymers determined by ^{13}C-NMR, it is possible to calculate the parameters $\overline{n}_E$ and $\overline{n}_P$, which represent the relative average number of ethylene and propylene units in the copolymer chain, respectively, and they give information about the distribution of comonomers along the copolymer chain. The values calculated for $\overline{n}_E$ and $\overline{n}_P$ are also listed in Table 2 and they can be related directly to the crystallinity obtained by DSC analysis. It can be concluded that, in the heterogeneous copolymers in the conditions of the reactions, when a chain contains 5 units of the same consecutive monomer, it crystallizes.

Another set of experimental properties related to crystallinity is the elution temperature (T_e) and the percentage of soluble fraction (SF%) determined by TREF.

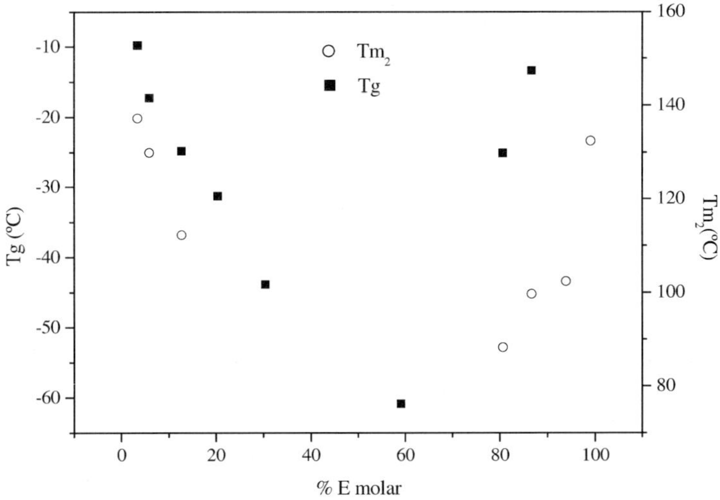

Figure 3.
Glass transition and second melting temperatures for the copolymers in function of the ethylene content (% E molar).

Figure 4 shows the results obtained by TREF as a function of the ethylene comonomer content and Figure 5 shows the TREF profiles for these copolymers.

This technique was limited to copolymers with very different comonomer contents, since in the middle range of comonomer content an amorphous copolymer was reached. In this case, the limit is lower than 10% molar of ethylene and higher than 80% molar of ethylene. For copolymers with low ethylene content (EP3 to EP13), values for T_e decreases faster with % E molar than in the high ethylene content (EP81 to EP99). This should be due to the fact that the crystallinity of copolymers with high propylene content is influenced by the composition and sequence distribution, besides tacticity. This make the study of the chemical composition distribution of polypropylene copolymers more complicated. The study of both tacticity and composition distribution should be completed with other techniques like FTIR. For low and high comonomer contents, the *SF* is low too, while for intermediate comonomer contents the *SF* is very high, as expected.

Table 2.
Crystallinity and $\bar{n}_E$ and $\bar{n}_P$ values for the copolymers.

Copolymer	Crystallinity (%)[a]	$\bar{n}_E$	$\bar{n}_P$
EP3	38.9	1.05	30.06
EP6	35.3	1.20	20.42
EP13	19.0	1.15	7.88
EP21	–	1.14	4.51
EP31	–	1.25	2.85
EP59	–	2.07	1.41
EP81	18.1	4.64	1.14
EP87	28.2	6.67	1.12
EP94	37.5	15.00	1.10
EP99	51.1	81.84	1.00

[a] Calculated from the second melting point.

In Table 3 the weight average molecular weight ($\overline{M}_w$) and the polydispersity index ($\overline{M}_w/\overline{M}_n$) of the copolymers by GPC and GPC-MALS are shown.

$\overline{M}_w$ and $\overline{M}_w/\overline{M}_n$ values obtained by GPC are higher than the ones obtained by GPC-MALS. The $\overline{M}_w$ values are dependent on comonomer content, showing the lowest values from 20 to 60%E molar. The polydispersity index is approximately constant, as expected for this single-site catalyst.[28]

The use of a multiangle light scattering detector allows the evaluation of both the molecular weight and the mean radius of gyration (R_g), for each slice of the size exclusion chromatogram. The dependence

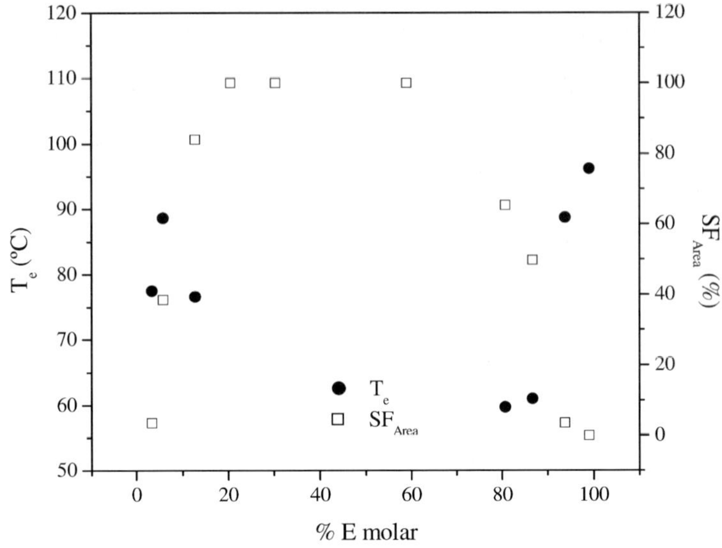

Figure 4.
Influence of comonomer content on TREF elution temperature and percentage soluble fraction.

of R_g on molecular weight is established by the scaling law $R_g = Q \cdot M_w^q$.[29] Figure 6 shows the plot of R_g *versus* M_w obtained for the copolymers that were shown in Figure 1, together with values for polyethylene (NBS 1475) and polypropylene (PP isotactic), taken as reference.

The slope of the plot provides the q coefficient which is related with the shape of the chain. The results obtained for the parameter q for all the samples studied are shown in Table 4. Values of q in the range 0.5 to 0.6 are typical for linear random-coil chains.[30]

A value of $q = 0.54$ was obtained for both homopolymers, thus showing a typical random coil in 1,2,4-TCB. However, the presence of comonomer decreases the value

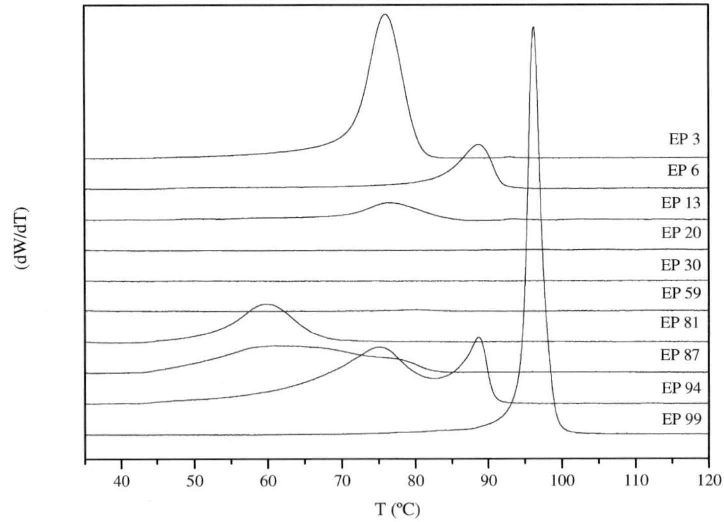

Figure 5.
Influence of comonomer content on TREF profiles.

Table 3.
Molecular weight and polydispersity index obtained by GPC and GPC-MALS.

Samples	GPC		GPC-MALS	
	$\overline{M}_w$ (g/mol)	$\overline{M}_w/\overline{M}_n$	$\overline{M}_w$ (g/mol)	$\overline{M}_w/\overline{M}_n$
EP3	112387	2.049	95 900	1.664
EP6	115940	2.215	98 270	1.780
EP13	86273	2.092	73 490	1.658
EP21	71163	2.005	57 450	1.687
EP31	61636	2.084	49 360	1.691
EP59	68752	2.289	54 470	2.001
EP81	128611	2.760	98 910	2.266
EP87	151595	2.647	11 8500	2.213
EP94	268488	2.780	20 1500	2.361
EP99	303465	3.382	22 7000	2.530

Table 4.
Values of q for ethylene/propylene copolymers.

Samples	q
PP[a)]	0.54
EP3	0.51
EP6	0.50
EP13	0.47
EP21	0.47
EP31	0.42
EP59	0.47
EP81	0.47
EP87	0.50
EP94	0.50
EP99	0.53
NBS 1475[b)]	0.54

a) polypropylene isotactic (Sigma-Aldrich);
b) polyethylene linear reference (NBS).

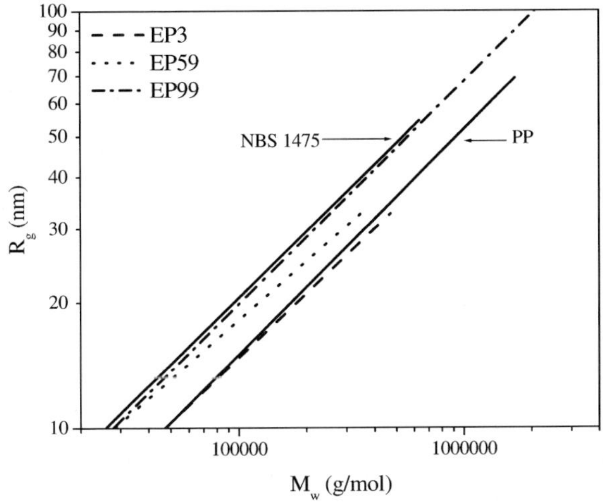

Figure 6.
Relationship between radius of gyration and molecular weight for some copolymers.

for the parameter q, reaching a value of 0.42 for copolymers with an ethylene molar percentage of 31%. This result shows how the presence of comonomers modifies the random coil structure of a homopolymer in the same way as the presence of long chain branching.

Conclusions

Good control system for the gas phase C_2/C_3 ratio in the polymerization process allows obtaining copolymers in the whole composition range. Different studies in function of composition have been carried out. SEM clearly shows the lack of morphology in copolymers close to 50/50 ethylene/propylene ratio, which could be related to the absence of crystallinity of those samples. DSC and XRD analyses show, for copolymers with low comonomer content, similar behaviour to that of the corresponding homopolymer, but when the comonomer content increases, the melting temperature and glass transition temperature decrease and the copolymer becomes amorphous. TREF analysis indicates that copolymers containing between 10% and 80% of ethylene comonomer are essentially amorphous, while copolymers with ethylene comonomer content out of this range are semicrystalline in the tested conditions.

GPC-MALS technique shows the dependence of M_w with comonomer content, showing a minimum in the intermediate position. The incorporation of comonomer also affects the shape (random coil) of the polymer in solution.

Acknowledgements: The authors grateful acknowledge REPSOL-YPF. MEC (PPQ2000-042-P4-02) and CAM (URJC/CAM 05-06. URJC-MYN-034-1).

[1] O. P. Grigoryeva, A. M. Fainleib, A. L. Tolstov, O. M. Starostenko, E. Lievana, J. Karger-Kocsis, *Journal of Applied Polymer Science* **2005**, *95*, 659.
[2] M. Pires, R. Mauler, S. Liberman, *Journal of Applied Polymer Science* **2004**, *92*, 2155.
[3] J. Voegele, C. Troll, B. Rieger, *Macromol. Chem. Phys.* **2002**, *203*, 1918.
[4] D. Arrowsmith, W. Kaminsky, A.-M. Schauwienold, U. Weingarten, *Journal of Molecular Catalysis A: Chemical.* **2000**, *160*, 97.
[5] M. Galimberti, F. Piemontesi, O. Fusco, I. Camurati, M. Destro, *Macromolecules* **1998**, *31*, 3409.
[6] A. Ioku, T. Shiono, T. Ikeda, *Appl. Catal. A.* **2002**, *226*, 15.
[7] M. R. Ribeiro, A. Deffieux, M. F. Protela, *Ind. Eng. Chem.* **1997**, *36*, 1224.
[8] S. Liu, F. Meng, G. Yu, B. Huang, *Journal of Applied Polymer Science* **1999**, *71*, 2253.
[9] A. Quijada, R. Rojas, L. Alzamora, J. Retuert, F. M. Rabagliati, *Catal. Lett.* **1997**, *46*, 107.
[10] G. Hlatky, *Chem. Rev.* **2000**, *100*, 1347.
[11] G. Fink, B. Steinmetz, J. Zechlin, C. Przybyla, B. Tesche, *Chem. Rev.* **2000**, *100*, 1347.
[12] G. Guerra, G. Maurizio, F. Piemontesi, O. Ruiz de Ballesteros, *J. Am. Chem. Soc.* **2002**, *124*, 1566.
[13] R. Medrano, M. T. R. Laguna, E. Saiz, M. P. Tarazona, *Phys. Chem. Chem. Phys.* **2003**, *5*, 151.
[14] M. Neotopilik, M. Bohdanecký, P. Kratochvíl, *Macromolecules* **1996**, *29*, 6023.
[15] W. Radke, P. F. W. Simon, A. H. E. Müller, *Macromolecules* **1996**, *29*, 4926.
[16] R. van Grieken, G. Calleja, D. Serrano, C. Martos, A. Melgares, I. Suarez, *Polym. Reac. Eng.* **2003**, *11*, 17.
[17] J. C. Randall, *J. Macromol. Sci. Part C: Rev. Macromol. Chem. Phys.* **1989**, *29*, 201.
[18] M. Kakugo, *Macromolecules* **1982**, *15*, 1150.
[19] B. H. Zimm, W. Stockmayer, *The Journal of Chemical Physics* **1949**, *17*, 1301.
[20] P. J. Wyatt, *Analytica Chimica Acta.* **1993**, *272*, 1.
[21] B. Coto, J. M. Escola, I. Suárez, M. J. Caballero, *Polymer Testing.* **2007**. In press.
[22] G. Weickert, G. B. Meier, J. T. M. Pater, K. R. Westerterp, *Chem. Eng. Sci.* **1999**, *54*, 3291.
[23] M. Carcuchinski, C. Krug, J. Dupont, G. Barrera, J. Zimnoch, T. Uozumi, T. Sano, K. Soga, *Journal of Molecular Catalysis A: Chemical* **2001**, *169*, 275.
[24] R. Bouza, C. Marco, Z. Martín, M. A. Gómez, G. Ellis, L. Barral, *Journal of Applied Polymer Science* **2006**, *102*, 6028.
[25] Z. Ali, Y. Badr, W. Eisa, *Polymer Composites* **2006**, *10*, 709.
[26] G. Akovali, A. Atalay, *Polymer Testing* **1997**, *16*, 165.
[27] H. Dweik, A. Al-Jabareen, G. Marom, E. Assouline, *Internacional Journal of Polymeric Materials* **2003**, *52*, 655.
[28] D. Bianchini, K. M. Bichinho, J. H. Z. dos Santos, *Polymer* **2002**, *43*, (10), 2937.
[29] J. Búrdalo, R. Medrano, E. Saiz, M. P. Tarazona, *Polymer* **2000**, *41*, 1615.
[30] S. Podzimek, T. Vlcek, *Journal of Applied Polymer Science* **2001**, *82*, 454.

Characterization of Polyethylene Nascent Powders Synthesized with $TpTiCl_2(OR)$ Catalysts

Emilio Casas,[1] *Arquímedes Karam,**[1] *Antonio Díaz-Barrios,*[1] *Carmen Albano,*[1] *Yanixia Sánchez,*[1] *Bernardo Méndez*[2]

Summary: Different kinds of polyethylene and ethylene-1-hexene copolymers were synthesized with $TpTiCl_2(OR)$ (Tp = hydrotris(pyrazolyl)borate; R = Et, i-Pr, n-Bu) catalysts with and without H_2. The polymers were characterized by ^{13}C NMR, capillary viscosimetry or GPC, and DSC. The homopolymers showed properties characteristic of ultra-high molecular weight polyethylenes (UHMWPE) with linear structure and high density polyethylenes (HDPE) with molecular weights in the range of commercial grades under hydrogen atmosphere. The copolymers showed a 1-hexene incorporation up to 6 mol-%. Important differences in the thermal properties were observed between the first DSC (nascent powders) and the second DSC heatings (melt-crystallized samples), which evidenced the molecular weights influence on the melt-crystallized samples.

Keywords: hydrogen transfer; 1-hexene; nascent powders; polyethylene; $TpTiCl_2(OR)$

Introduction

Ethylene can be polymerized under various conditions to yield polyethylenes having markedly different chain structures and physical properties as a function of different structural parameters that influence their ultimate properties, such as type, amount and distribution of comonomer(s) and branching, average molecular weight (MW) and molecular weight distribution (MWD).[1] In the last few years, authors have pointed out that polyethylene reactor powders have a unique morphology and higher crystallinity than melt-crystallized or solution-crystallized samples.[2–4]

Better understanding of the nature of the nascent morphology of polyolefins, namely the formation and molecular organization of polymer particles initiated by heterogeneous or homogeneous catalyst systems in the reactor, and its related physical properties, is an important research topic.[2,5] The development of the nascent state morphology of polyolefins in the reactor is reasonably well understood on the micrometer level. However, at the molecular scale, the physical properties of nascent polyolefins are less clearly understood.[6,7] Although different catalyst types and synthesis conditions clearly play a role, there is no consensus on the origin of the high value for the peak melting point commonly seen in nascent polyethylene.[2,8–16] Interpretations which invoke chain-extended and/or fibrilar crystals,[8–11] strained noncrystalline tie points,[12,13] instrumental effects,[2] or small crystals which reorganize on heating,[14,15] have been invoked to describe aspects of the thermal behavior. Nascent samples of UHMWPE also exhibit a peak melting point which approaches the equilibrium value (T_m^0) for polyethylene, but is irreversibly lowered following a melt crystallization cycle.[2,10,14,16] In this sense, the nascent morphology of UHMWPE represents a unique combination of morphology, crystallinity and melting characteristics

[1] Laboratorio de Polímeros, Centro de Química, Instituto Venezolano de Investigaciones Científicas (IVIC), Apartado 21827, Caracas 1020-A, Venezuela
Tel.: +58-212-5041638, Fax: +58-212-5041350
E-mail: akaram@ivic.ve

[2] Facultad de Ciencias, Escuela de Química, Universidad Central de Venezuela

which is generally inaccessible via melt crystallization of polyethylene.[16]

Although the high melting temperature of nascent polyolefins is known for more than three decades, the thermal properties of polyethylenes made with non-metallocene precatalysts have not been investigated. The alkoxyl effect of $TpTiCl_2(OR)$ (R = Et, *i*-Pr, *n*-Bu) complexes on the polymerization of ethylene,[17,18] ethylene-1-hexene, and ethylene in the presence of H_2 have been recently investigated in our laboratories. The aim of the present work is the characterization the polymers obtained by these complexes in order to compare the physical properties of polyethylenes and the differences between nascent powders and melt-crystallized samples.

Experimental Part

The polymers were characterized by ^{13}C NMR using a JEOL 270 Spectrometer. The polymer samples were prepared in tetrachloroethane or 1,2,4-trichlorobenzene (TCB) in a 5 mm NMR tube. The 1-hexene content was calculated according to the ASTM X70-8605-2 method.[19]

Average molecular weights were measured in TCB with 0.2% v/v of IRGANOX® at 135 °C using a high temperature gel permeation chromatographer (GPC). The average molecular weights and the polydispersities were obtained using a universal calibration curve made with polystyrene standards. When the polymer was not dissolved in TCB, the viscosity-average molecular weights were established by one-point determination of the intrinsic viscosity measured from an Ubbehlode dilution viscometer in decalin stabilized with 0.2% v/v of IRGANOX® at 135 °C. The average viscosimetric molecular weights were determined using the Schulz-Blaschke correlation.[20]

The thermal properties of the polymers were analyzed in a METTLER TOLEDO, DSC822e calorimeter. The essays were carried out using the following procedure: polyethylene samples (3 to 10 mg) were sealed in an aluminium pan, heated at 10 °C/min from 25 °C to 170 °C (1st heating), kept 5 minutes at 170 °C isothermally, cooled at 10 °C/min from 170 °C to 25 °C, and finally heated at 10 °C/min from 25 °C to 170 °C (2nd heating). The analysis of the final curves led to the melting peak temperature (T_m) and the melting enthalpy (ΔH). The crystallinity degrees (X_c) were calculated using the expression $(\Delta H_{sample}/\Delta H_{theoric}) \times 100$, where ΔH_{sample} is the experimental value and $\Delta H_{theoric}$ is the melt enthalpy of crystalline polyethylene (290 J/g).[21] Additionally, the superheating phenomena was tested varying the sample mass, the first heating rate, and the temperature of isothermal treatment when the thermal history was erased.

Results and Discussions

The polyethylenes were characterized by ^{13}C NMR, capillary viscosimetry measurements and DSC. The ^{13}C NMR analysis of the polyethylenes showed a single signal of methylene group (CH_2) at 29.59 ppm. Signals corresponding to terminal methyl groups (CH_3) and tertiary carbon (CH) were not detected due to their relatively low concentrations with respect to the methylene groups (CH_2), which is an evidence that all polymers were mainly linear. Figure 1 shows the ^{13}C NMR spectra of the polyethylene made with $TpTiCl_2(OEt)$, which is representative of the all polyethylenes evaluated.

The capillary viscosimetry analysis showed that the average viscosimetric molecular weights (M_v) were strongly influenced by the alkoxyl ligands. As can be seen in Table 1, the M_v of the polymers increases as the $TpTiCl_2(OR)$ steric hindrance increases showing the following tendency: *i*-Pr > *n*-Bu > Et. This result suggests that the alkyl chain size reduces the ratio of chain transfer rate to propagation rate. This property allowed us to classify the samples as UHMWPE. The DSC analysis of the polyethylenes showed melting

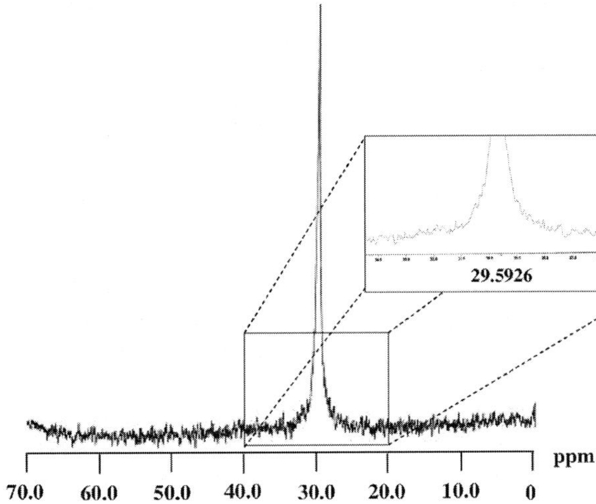

Figure 1.
^{13}C NMR of polyethylene produce by $TpTiCl_2(OEt)$.

temperatures around 138 °C and crystallinities between 39 to 47% (in the second heating), which agrees with UHMWPE characteristics.

The thermal properties between the nascent powders (1st heating) and the melt-crystallized samples (2nd heating) of these polyethylenes were significantly different. Although the melting temperature does not change significantly with the type of alkoxyl group used in the catalyst, there is a difference of almost 8 °C between the first and second heating. The nascent powder samples exhibit a high peak melting temperature at 145 °C, while the melt-crystallized samples show a lower peak melting temperature around 138 °C. This behavior might indicate that the crystals in the nascent polymer are better formed (larger) than the crystals formed after the first melt, which implies that the crystallization kinetics of nascent and molten polyethylene are different. However, superheating phenomena could possibly be present due to heat diffusion problems in the samples. Similar trends were found when comparing the crystallinities of the nascent polymers and melt-crystallized samples. In all cases, the crystallinities of the nascent powders were higher than those of the melt-crystallized samples, reaching values up to 76% and 47%, respectively.

Table 1.
Physical properties of polyethylenes.

Catalyst	$M_v \times 10^5$ (g/mol)	1st Heating		2nd Heating	
		T_m (°C)	X_c (%)	T_m (°C)	X_c (%)
$TpTiCl_2(OEt)$	23.3	145.8	69	137.7	39
$TpTiCl_2(OiPr)$	34.8	145.7	76	137.9	47
$TpTiCl_2(OnBu)$	25.6	145.2	75	138.0	46

Figure 2, 3 and 4 show the effect of sample mass (3, 5, 7 and 9 mg.), first heating rate (2, 5 and 10 °C/min), and the temperature of isothermal treatment when the thermal history was erased (170, 190, 210 and 227 °C) on the melting temperature and crystallitinity of the first and second heating for polyethylene made with $TpTiCl_2(OEt)$. As shown in Figure 2.a, the melting temperature of the first and second heating increases slightly (around 1 to 2 °C) with sample mass, remaining constant between 3 to 7 mg. This trend is in agreement with results reported the literature,[8–10] where superheating occurs when the sample mass increases. However, the differences in melting temperatures between the 1st and 2nd meltings when different polymer masses were used were the same in all cases

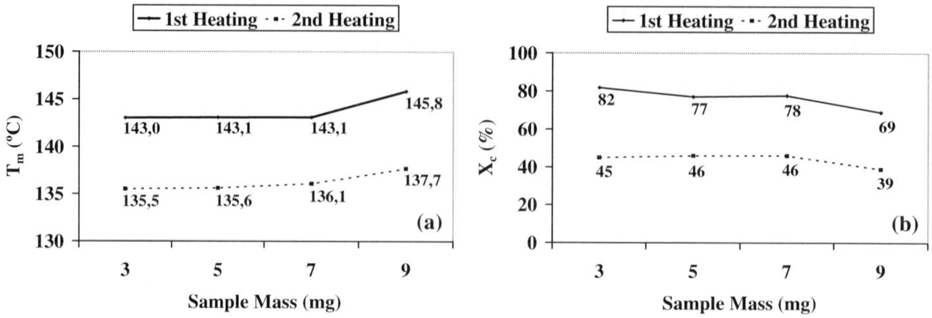

Figure 2.
Effect of sample mass on melting temperature (a) and crystallinity (b).

(around 8 °C). In contrast, the crystallinity presented an opposite behaviour to the melting temperature (Figure 2.b). It was observed that the crystallinity increases when the sample mass decreases. In agreement with other authors,[22] when the polymer is treated above the melting temperature, the entanglement density increases with the mass, reducing the chain mobility and hindering the crystallization. Despite of this fact, the differences of crystallinities between 1st and 2nd heatings for samples of different masses were the same in all cases (around 30%).

We also evaluated the influence of the 1st heating (thermal history erasing) at different rates (2, 5 and 10 °C/min) on these properties. As can be observed in Figure 3.a, the melting temperature increases around 4 °C between 5–10 °C/min (1st heating) with the heating rate. However, significant variations were not observed in the 2nd heating (2 °C approximately). In contrast, for both the 1st and 2nd heatings the crystallinity decreases when the heating rate increases (Figure 3.b), showing a maximum at 5 °C/min (45% and 79%, respectively). The differences in crystallinity were around 30% between each heating of each thermal history erasing rate tested. This behaviour is similar to the described in the sample mass case.

Some authors[23,24] have proposed that UHMWPE exhibits high concentration of tie molecules. When the nascent UHMWPE aggregates are heated up to 190 °C, the polymers retains its crystalline entities due to the long molecules entanglement points, which are present even after melting.[25,26] In this sense, the effect of the temperature of the isothermal treatment was analysed upon the melting temperature and crystallinity, as shown in Figure 4. Figure 4.a shows that, in the 1st heating, the melting temperature does not show a significant variation when the temperature of the

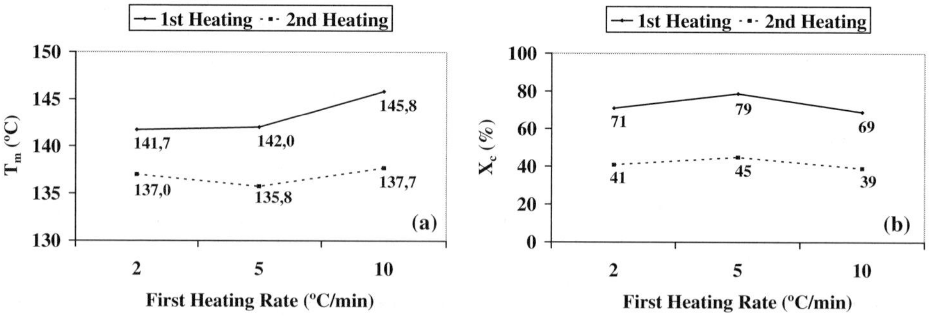

Figure 3.
Effect of first heating rate on melting temperature (a) and crystallinity (b).

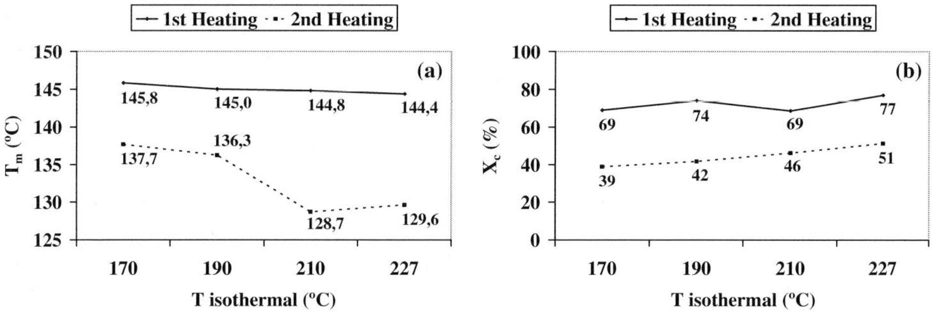

Figure 4.
Effect of isothermal temperature on melting temperature (a) and crystallinity (b).

isothermal treatment was increased. However, in the second heating, the melting temperature evidenced a significant drop (around 7 °C) when the nascent powder was heated over 190 °C. This behaviour might indicate that all crystalline entities initially present in the polyethylene were destroyed above 190 °C. Regarding the crystallinity, both 1st and 2nd heatings showed that the crystallinity increases with the temperature of isothermal treatment. In agreement with the literature,[23,26] this tend suggest that the tie molecules were completely destroyed during the melting process.

These experimental results indicated that the best DSC conditions for this kind of polyethylenes were: sample mass = 7 mg, heating rate = 10 °C/min and isothermal temperature treatment = 170 C, which were used to evaluate the polymers obtained by the other systems.

For the $TpTiCl_2(OR)/MAO$/ethylene/1-hexene system, the copolymers were characterized by ^{13}C NMR, GPC and DSC. The

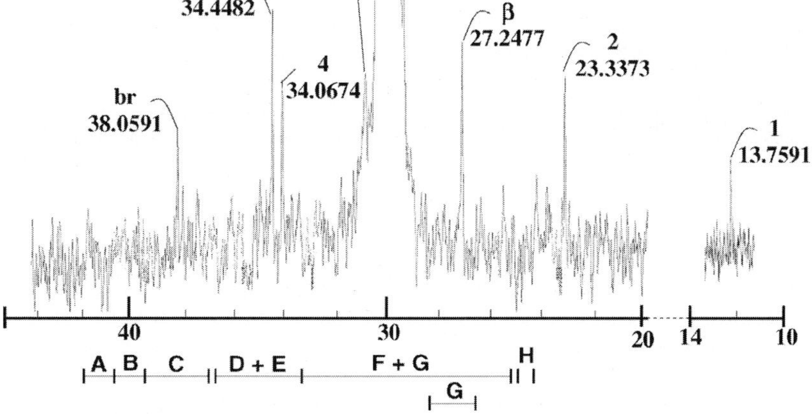

Figure 5.
NMR ^{13}C of copolymer made with $TpTiCl_2OEt$ complex ([1-Hex] = 0.5)

Table 2.
Physical properties of ethylene-1-hexene copolymers.

Catalyst	[1-Hex] (M)	C_6 mol-(%)	$M_w \times 10^5$ (g/mol)	M_w/M_n	1st Heating T_m (°C)	1st Heating X_c (%)	2nd Heating T_m (°C)	2nd Heating X_c (%)
	0.3	3.3	18.7	28	136.7	59	130.2	48
$TpTiCl_2(OEt)$	0.4	2.2	14.8	27	137.2	59	130.7	47
	0.5	6.1	10.4	25	134.2	60	126.8	49
$TpTiCl_2(O\textit{i}Pr)$	0.5	1.2	6.2	21	135.7	58	128.8	47
$TpTiCl_2(O\textit{n}Bu)$	0.5	3.4	16.4	33	137.0	64	129.5	47

presence of 1-hexene in the copolymer was analyzed by ^{13}C NMR. Figure 5 shows an spectrum of a copolymer made with $TpTiCl_2(OEt)$ which is representative of those made with $TpTiCl_2(OR)$ complexes. According to the ASTM X70-8605-2 method, all the spectra showed characteristic signals of ethylene/1-hexene copolymers.

As seen in Table 2, the 1-hexene content in the copolymers synthesized with $TpTiCl_2(OEt)$ for the established comonomer concentrations showed a higher value at 0.5 M (6.1 mol %). The lower 1-hexene content (1.2 mol %) was obtained with $TpTiCl_2(O\textit{i}Pr)$, which was attributed to catalyst steric effects. The average molecular weights of the copolymers made with $TpTiCl_2(OEt)$ decreases from 18.7×10^5 to 10.4×10^5 g/mol as the 1-hexene concentration in the feed was increased (0.3–0.5 M). These results are in agreement with the propagation rate of comonomer insertion and 1-hexene concentration. The copolymers showed wide molecular weight distributions, as can be noted in Table 2, in agreement with the results obtained by Nakazawa et al.,[27] Jordan et al.,[28–30] and our previous study[18] on ethylene polymerization. This fact is an evidence that more of one active species is present in the catalyst.

On the other hand, Table 2 shows the 1-hexene concentration effect on the 2nd melting temperature. The melting temperature did not vary significantly when the 1-hexene concentration in the feed was between 0.3–0.4. However, at the maximum 1-hexene concentration in the feed (0.5), the melting temperature drops down. This behavior might indicate that when the 1-hexene content in the copolymer increases, the side chains hinder the growth of crystal lamellae, bringing about a drop of melting temperature. The copolymers showed melting temperatures ranging from 126 to 131 °C and crystallinities between 47 to 49 %. Additionally, the effect of alkoxyl group type on thermal properties was not significant. All these properties allowed to infer that the obtained copolymers showed high molecular weights (around 10×10^5 g/mol) with low comonomer incorporation (up to 6 mol-%).

A difference of 15% was observed when the thermal properties of the nascent powders (1st heating) and the melt-crystallized sample (2nd heating) of the copolymers were compared. A similar behavior was found in the homopolymers (Table 1), which showed a difference of 30% in its crystallinities. The lower difference observed in the copolymers case can be a consequence of the entanglement density reduction as a result of lower molecular weights present in the copolymers. This behavior allows higher chain mobility during the crystallization process.

For the $TpTiCl_2(OR)$/MAO/ethylene/H_2 system, the homopolymers were characterized by ^{13}C NMR, GPC and DSC. The results are summarized in Table 3.

The ^{13}C NMR analysis of the polyethylene samples showed a single signal of methylene group at 29.5 ppm, which shows that all polymers were linear. It is worthy mentioning that the weight average molecular weights (M_w) were strongly influenced

Table 3.
Physical properties of polyethylenes made in the presence of H_2.

Catalyst	$P_{ethyl/hydrogen}$	$M_w \times 10^5$ (g/mol)	M_w/M_n	1st Heating		2nd Heating	
	(-)			T_m	X_c	T_m	X_c
				(°C)	(%)	(°C)	(%)
	10: 1	5.23	10.2	141.2	72	139.0	67
$TpTiCl_2(OEt)$	7: 1	2.24	6.3	141.7	68	140.1	65
	5: 1	1.04	9.3	137.6	77	138.4	75
$TpTiCl_2(O\mathit{i}Pr)$	5: 1	0.72	4.8	140.2	80	138.7	73
$TpTiCl_2(O\mathit{n}Bu)$	5: 1	0.58	8.5	138.9	62	138.1	67

by hydrogen pressure (Table 3). Thus, when the $P_{ethylene/hydrogen}$ ratio was increased, the M_w increased from 1.04×10^5 to 5.23×10^5 g/mol. This tendency was expected in accordance with industrial practice, where the decrease of the hydrogen pressure reduces the probability of the transfer reaction to occur as a consequence of the drop of the hydrogen concentration in the reaction medium around the active site. On the other hand, the polyethylene polydispersities were lower than the copolymers case. This might indicate that, for the catalytic systems evaluated, hydrogen is capable of inhibiting a fraction of active species during the polymerization.

The polyethylenes showed melting temperatures in the interval from 138 to 140 °C and crystallinities between 65 to 75%. Additionally, no significant effects on thermal properties were observed when the type of alkoxyl group was changed. The characterization results evidenced that all the properties corresponded to a high density polyethylene (HDPE) with molecular weights in the range of commercials grades.

As seen in Table 3, it was found that the melting temperature and crystallinity did not show significant variation between the 1st and 2nd heatings. This behaviour shows that molecular weight has an important influence on the thermal properties of the melt-crystallized sample: the differences in the thermal properties between the nascent powders and the melt-crystallized samples were higher when the molecular weight increased.

Conclusions

The characterization by ^{13}C NMR, capillary viscosimetry or GPC, and DSC, showed that the catalytic system has an important influence on the final properties of nascent powders. Important differences were observed between the thermal properties obtained from the first DSC heating (nascent powders) and the second DSC heating (melt-crystallized samples). This behaviour was attributed to the crystallization process during the polymer synthesis which likely leads to less entangled morphologies (better formed crystals) than those obtained with the melt-crystallized sample, due to the fact that the crystallization kinetics of the nascent and molten polyethylene are different. Furthermore, an important influence of the molecular weight on the melting temperature and crystallinity of the melt-crystallized samples was corroborated.

Acknowledgements: We acknowledge the Chemistry Center of Venezuelan Institute for Scientific Research (IVIC) for financial support for this research and FONACIT for Grants S1-2000000519 and F-2000001365.

[1] H. Knuuttila, A. Lehtinen, A. Nummila-Pakarinen, *Adv. Polym. Sci.* **2004**, *169*, 13.
[2] S. Ottani, R. S. Porter, *J. Polym. Sci.: Polym. Phys. B* **1991**, *29*, 1179.
[3] Y. M. T. Engelen, P. J. Lemstra, *Polym. Comm.* **1991**, 32, 3433.

[4] D. Hofmann, E. Schulz, D. Fanter, H. Fuhrmann, D. Bilda, *J. Appl. Polym. Sci.* **1991**, *42*, 863.
[5] J. Loos, M. Arndt-Rosenau, U. Weingarten, W. Kaminsky, P. J. Lemstra, *Polym. Bull.* **2002**, *48*, 191.
[6] S. Ottani, E. Ferracini, A. Ferrero, V. Malta, R. Porter, *Macromolecules* **1995**, *28*, 2411.
[7] R. A. Phillips, *J. Polym. Sci. Polym. Phys. B* **1998**, *36*, 495.
[8] H. D. Chanzy, R. H. Marchessault, *Polymer* **1967**, *8*, 567.
[9] H. D. Chanzy, J. F. Revol, R. H. Marchessault, H. Lamandé, *Kolloid Z. Z. Polym.* **1973**, *251*, 563.
[10] H. D. Chanzy, E. Bonjour, R. H. Marchessault, *Kolloid Z. Z. Polym.* **1974**, *252*, 8.
[11] E. Hellmuth, B. Wunderlich, *J. Appl. Phys.* **1965**, *36*, 3039.
[12] G. Delmas, *J. Polym. Sci. Polym. Phys. Edn.* **1993**, *31*, 2011.
[13] F. G. Morin, G. Sdelmas, D. F. R. Gilson, *Macromolecules*, **1995**, *28*, 3248.
[14] Y. M. T. Tervoort-Engelen, P. J. Lemstra, *Polym. Comm.* **1991**, *32*, 343.
[15] S. Rastogi, L. Kurelec, P. J. Lemstra, *Macromolecules* **1998**, *31*, 5022.
[16] R. A. Phillips, *J. J. Polym. Sci. Part B: Polym. Phys.* **1998**, *36*, 495.
[17] Karam, M. Jimeno, J. Lezama, E. Catarí, A. Figueroa, B. Rojas de Gascue, *Journal of Molecular Catalysis A: Chemical* **2001**, *176*, 65.
[18] Karam, E. Casas, E. Catarí, S. Pekerar, A. Albornoz, B. Méndez, *J. Mol. Catal. A: Chem.* **2005**, *238*, 233.
[19] M. De Pooter, P. B. Smith, K. K. Dohrer, K. F. Bennett, M. D. Meadows, C. G. Smith, H. P. Schouwenaars, R. A. Geerards, *Journal of Applied Polymer Science* **1991**, *42*, 399.
[20] *Polymer Handbook*, J. Brandrup, E. H. Immergut, Eds., 2nd edition, John Wiley & Sons, New York **1975**, p. V-14.
[21] J. G. Fatuo, *Morphology and Crystallization in Polyolefins*, Eds., C. Vasile, R. B. Seymour, *Handbook of Polyolefins*, **1993**, Marcel Dekker, Chapter 8.
[22] P. Smith, H. D. Chanzy, B. P. Rotzinger, *J. Mat. Scie.* **1987**, *22*, 523.
[23] L. Minkova, *Colloid Polym. Sci.* **1994**, *272*, 115.
[24] M. Mihailov, L. Minkova, E. Nedkov, R. Kircheva, *Makromol. Chem.* **1979**, *180*, 2351.
[25] M. Kresteva, E. Nedkov, A. Radilova, *Colloid Polym. Sci.* **1985**, *263*, 273.
[26] Z. Bashir, A. Keller, *Colloid Polym. Sci.* **1989**, *267*, 116.
[27] H. Nakazawa, S. Ikai, K. Imaoka, Y. Kai, T. Yano, *Journal of Molecular Catalysis A: Chemical* **1998**, *132*, 33.
[28] K. Minchue, R. F. Jordan, *Organometallics* **2004**, *23*, 460.
[29] K. Michiue, R. F. Jordan, *Macromolecules* **2003**, *36*, 9707.
[30] S. Murtuza, O. Casagrande, R. F. Jordan, *Organometallics* **2002**, *21*, 1882.

Characterization of LDPE grafted with Diethylmaleate by Gamma Radiation: Application of FTIR, GPC and SSA Techniques

Y. Sánchez,[1] *C. Albano,*[*1,2] *R. Perera,*[*3] *A. Karam,*[1] *P. Silva*[4]

Summary: The grafting of diethyl maleate onto low-density polyethylene using gamma irradiation as initiator has been evaluated. The grafting degree was estimated by FTIR and with the use of a calibration curve. The functionalized polymers were characterized by DSC, MFI, SSA and GPC. An increase in the grafting degree with the radiation dose and the concentration of the functional monomer was found. Branching seemed to be the preferential reaction induced by radiation, causing an increase in average molecular weights with the consequently reduction of melt flow index values. The molecular segregation induced by the calorimetric treatment showed that grafting occurs preferentially through secondary carbon sequences.

Keywords: diethylmaleate; gamma radiation; grafting; GPC; polyethylene; SSA

Introduction

The enhanced compatibility properties of grafted polyethylenes have increased their interest due to their potential applications, especially in the improvement of interfacial adhesion in composites and polymers blends,[1] as a consequence of the presence of polar groups able to provide interaction sites for hydrogen or covalent bonding.[2–5]

Polymer grafting has been reported by extrusion and solution methods.[3,5–6] Using these techniques, several functional monomers with unsaturated polar groups have been tested,[2–11] e.g.: maleic anhydride, maleic acid, acrylic acid, maleic and fumaric esters, diethyl maleate, etc. These studies additionally have shown that other reactions can occur while the functionalization reaction takes place, such as monomer homopolymerization, crosslinking and coupling reactions. However, some of these reactions can be prevented using diethyl maleate, because in this case insertion is the prevailing mechanism.[4–5]

As it is well known, the grafting mechanism requires a radical initiator able to promote the functional monomer insertion.[4] In this sense, the study of the effects of gamma rays in polymers[12–14] has allowed proposing grafting by gamma radiation as a promising method for the insertion of functional groups into different polymer matrices, considering its advantages such as the high radical generation rate with the additional exclusion of chemical initiator agents.[15–17]

The study of grafting by gamma irradiation could be similar to other methodologies used, where the grafting degree is related to monomer and initiator concentration, chemical nature of the substrate, with the additional consideration of the radiation conditions. However, it is well known that exposition to γ-rays can induce polymer modifications[18–19] Therefore, the selection of appropriate grafting conditions is an important task.

[1] Laboratorio de Polímeros, Centro de Química, Instituto Venezolano de Investigaciones Científicas (IVIC)
E-mail: calbano@ivic.ve
[2] Escuela de Ingeniería Química, Facultad de Ingeniería, Universidad Central de Venezuela, Caracas
[3] Universidad Simón Bolívar. Departamento de Mecánica, Caracas, Venezuela
E-mail: rperera@usb.ve
[4] Centro de Física, Instituto Venezolano de Investigaciones Científicas (IVIC)

In this sense, interesting results in preliminary studies have been found in our laboratory[20] when diethyl maleate (DEM) was grafted onto low-density polyethylene (LDPE) using different radiation doses and varying the monomer concentration. In this work, the grafted polymer was characterized to determine the degree of DEM incorporation and the γ-ray dose effects on its properties.

Experimental Part

A commercial low-density polyethylene (LDPE) supplied by CORAMER, C.A., with a MFI of 4.06 dg/min, determined at 190 °C and 2.16 kg according to ASTM D-1238 standard procedure, was used. Solutions were prepared at 10% wt/vol using a blend of *cis* and *trans* decahydronaphtalene (Decalin 99%), supplied by Riedel de Haën, as solvent. Ethanol and *n*-hexane were employed as washing solvents. Diethyl maleate (DEM), manufactured by Aldrich Chemical Company Inc. was used as the functional monomer.

Solutions of LDPE and 30% DEM in decalin were prepared and irradiated with γ-rays from a ^{60}Co source in air at a dose rate of 4.8 kGy/h at room temperature. The Integral doses were 50, 100, 200 kGy. The grafting degree (GD), was determined by Fourier transform infrared spectroscopy (FTIR) in a Nicolet Magna-IR 560 spectrometer, measuring the ratio of absorption band areas at 1740 cm^{-1}, corresponding to the C=O of the DEM, and at 1460 cm^{-1} ($A^{-1}_{1740cm}/A^{-1}_{1460cm}$) a band characteristic of polyethylene. Relative areas of these absorption bands are proportional to the concentration of carbonyl groups (C=O) in the polymer. The molar concentration percent (% molar) was estimated using a calibration curve reported elsewhere,[10] which relates this area ratio with the molar concentration of DEM determined experimentally by ^{13}C NMR.

Samples were characterized by FTIR using a NICOLET Magna-IR 560 equipment. Spectra were recorded from compression-molded films at 190 °C. The unsaturation bands were followed, identifying the characteristic bands present in the grafted polymers and establishing the corresponding band area ratios, using the peak at 1460 cm^{-1} as the internal standard band for LDPE.

The average molecular weights and molecular weight distribution were measured by gel permeation chromatography (WATERS Alliance GPCV 2000) at 135 °C with 1,2,4-trichlorobenzene as mobile phase, stabilized with 0.25 g/l of butylated hydroxytoluene (BHT). The results were analyzed using a universal calibration curve based on polystyrene standards.

The distribution of melting points induced by the self-nucleation annealing technique (SSA) developed by Muller et al.[21] was recorded in a Mettler Toledo DSC 821. Once the thermal history was erased, the sample was heated at 10 °C/min up to the selected self-seeding and annealing temperature (T_s), where it was isothermally kept for 5 min and then cooled down to 25 °C. Then, the sample was heated again to a new self-seeding and annealing temperature, which was 5 °C lower than the previous T_s and held again for 5 min before cooling to 25 °C at 10 °C/min. This thermal treatment was repeated, being each T_s 5 °C lower than the previous one, until the minimum temperature selected was reached. Finally, the sample was heated at 10 °C/min up to 170 °C and its thermogram recorded.

Results and Discussion

Figure 1 evidences the feasibility of using gamma radiation as initiator in the functionalization reaction of LDPE with DEM. The GD, known as the molar concentration of the functional monomer inserted into the polymer chains, was found to be dependant on the radiation dose, increasing as higher doses were used. The functionalized polymers in this study achieved a maximum grafting degree of 0.34 molar % at 200 kGy. That means that the polymer needs higher radiation energies in order to produce

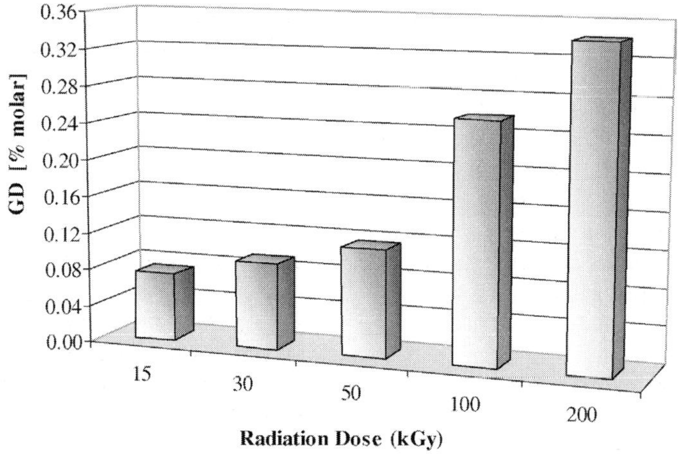

Figure 1.
Grafting degree of LDPE at different radiation doses, using 30% DEM in the solution.

enough number of radicals that can induce the functional monomer insertion.

FTIR spectra not only allowed identifying the DEM insertion, but also to identify[22] and quantify[6] the vinyl unsaturations, through the absorption bands analysis in the region 1000–800 cm^{-1}(Figure 2). The observed bands were assigned to out of plane C–H vibrations, associated to pendant vinyl groups ($R_2C{=}CH_2$) at 887 cm^{-1} and transvinylene groups (RCH=CHR) at 967 cm^{-1}. The presence of these bands depends on the radiation dose, because the transvinylene group is evident only when high radiation doses are employed. This group has been related[22] to the generation of long chain branching and additional unsaturations, due to the coupling reactions between radicals that take place, simultaneously, with the grafting reaction

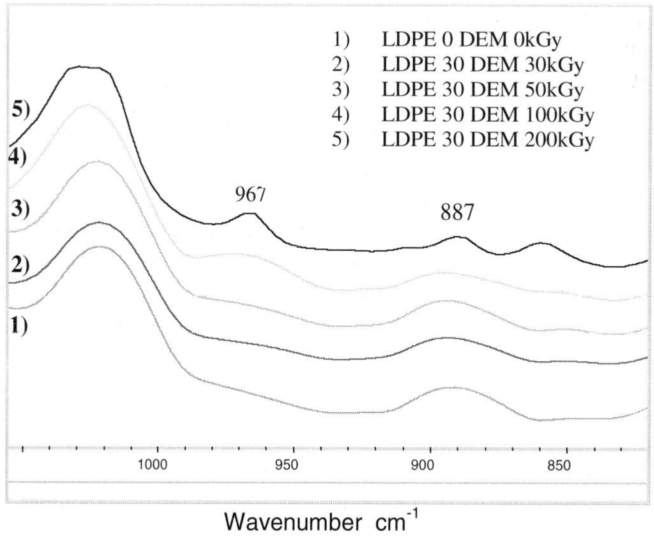

Figure 2.
FTIR of grafted LDPE at different radiation doses, using 30% DEM in the solution; detail of the 800–1000 cm^{-1} region.

(Figure 3). Theses bands were more pronounced at 200 kGy, evidencing that the branching reaction can occur additionally as a consequence of radiation effects.

Table 1 shows that the pendant vinyl groups are consumed during the grafting reaction, while the amount of *trans*-vinyl groups corresponding to long-branch formation increases, which is a consequence of coupling allylic radicals.[6] Additionally, the content of pendant vinyl groups increase with the radiation dose at doses higher than 50 kGy, which could be a consequence of other secondary reactions like branch scission, as Figure 4 shows. Branch scissions take place at lower proportions. Thus, their effect on the molecular-weight distribution curve is negligible. Additionally, the interconnection of the long branches could lead to crosslinking of the polymeric chains.

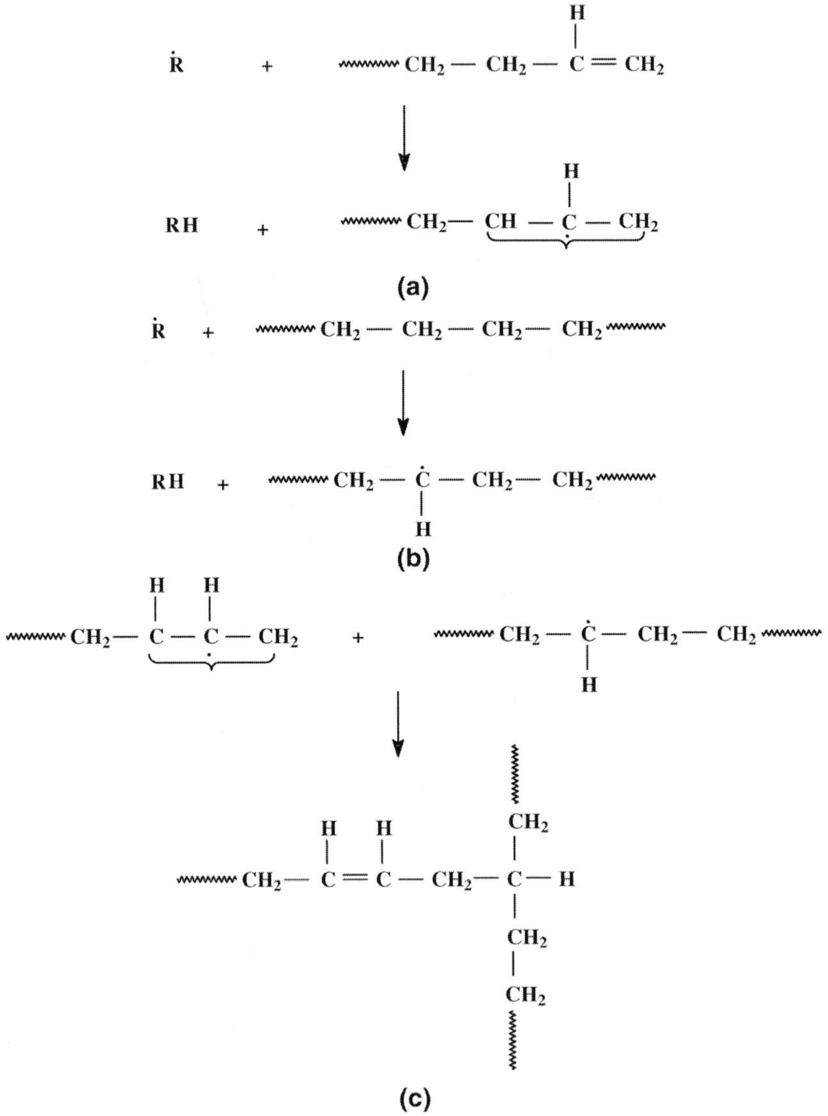

Figure 3.
Mechanism of long-chain branching and/or crosslinking: (a) allyl hydrogen abstraction; (b) H abstraction to form a secondary radical; (c) reaction of products from (a) and (b) to form long branches.[22]

Table 1.
Absorbance area ratios of unsaturations of grafted LDPE at different radiation doses, using 30% DEM in the solution.

	Absorbance Area Ratios of Unsaturations $\times 10^3$	
	Pendants Vinyls Group 887 cm^{-1}/1460 cm^{-1}	*Trans*-Vinyl Group 967 cm^{-1}/1460 cm^{-1}
0DEM_0kGy	8.9	0.0
30DEM_30kGy	2.7	0.2
30DEM_50kGy	2.8	0.2
30DEM_200kGy	3.5	4.7
30DEM_400kGy	6.7	10.6

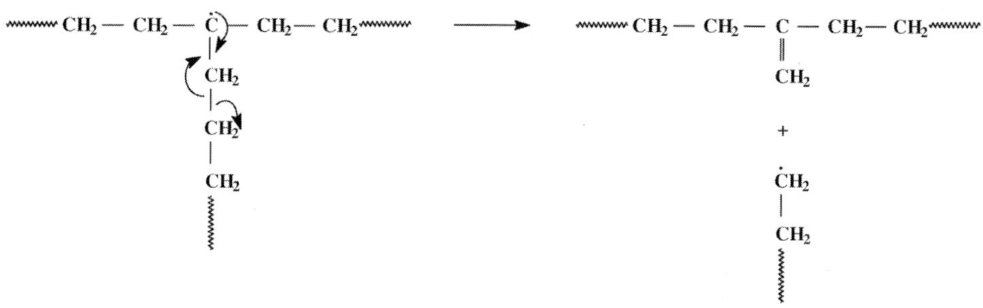

Figure 4.
Formation of pendant unsaturation in tertiary radical.[23]

Figure 5 shows the molecular weigh distribution curves of the grafted products, which show a noticeable displacement towards the fractions of higher molecular weights. An increase in the high molecular weight portion of the distribution curve with the radiation dose is clearly seen. This effect was attributed to the branching or crosslinking reactions that caused an increase in the average molecular weight compared to that of the unmodified LDPE, as shown in Table 2. The polydispersity

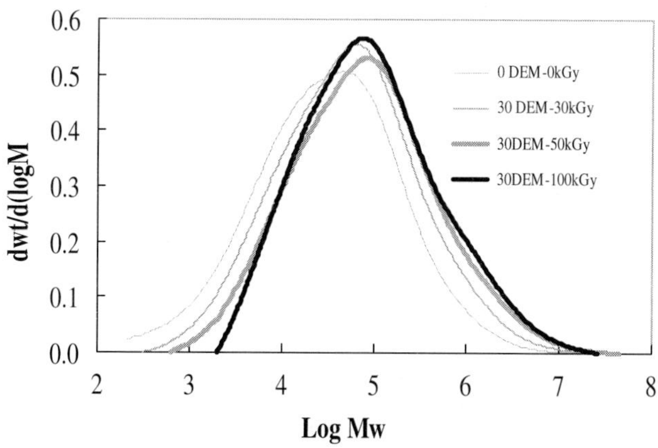

Figure 5.
Molecular-weight distribution curves of grafted LDPE at different radiation doses, using 30% DEM in the solution.

Table 2.
Average molecular weights of grafted LDPE at different radiation doses, using 30% DEM in the solution.

Radiation Dose	M_n	M_w	PDI
0DEM_0 kGy	8233	158680	19.3
30DEM_30 kGy	13099	239267	18.4
30DEM_50 kGy	20109	387837	19.3
30DEM_100 kGy	28982	410477	14.2

index (PDI) value change only at high radiation doses, where a reduction in its value was observed.

The final DSC curves, obtained after applying the SSA technique, showed the heterogeneity in the branching distribution, according to the segregation mechanism based on similar lengths of linear crystallizable sequences.[21] The final SSA thermograms showed nine fractions (Figure 6), corresponding to the segregation of the chains as a function of the branching content. As it is clearly noticed, the area under the peak with the highest melting temperature decreases significantly with the radiation dose, corresponding to the increase in grafting degree. This fact indicates that the insertion of DEM produced an interruption in the more linear sequences, showing that the grafting occurs preferentially in secondary carbons.[21,22,24–25] This fact could be corroborated comparing the results showed in Table 3, where a significant reduction in the partial area of the peak with the highest melting point in the grafted LDPE is observed, simultaneously with an increase in the melting peak areas of the fractions with higher branch contents. The maximum area reduction was observed in the LDPE grafted at 200 kGy. However, this modification could also be attributed to long-branch generation, as a consequence of radiation effects, which are as well able to interrupt the linear sequences. This fact was evidenced comparing the SSA curve of the neat irradiated LDPE, as shown in Figure 7. Additionally, no significant changes were observed in the melting temperatures of each fraction, which only depended on the chosen self-seeding temperature.

Conclusions

The characterization results evidenced the grafting of LDPE by gamma rays. LDPE needs high radiation doses in order to produce high grafting degrees. The generation of branching and unsaturations in the LDPE molecular structure depends on

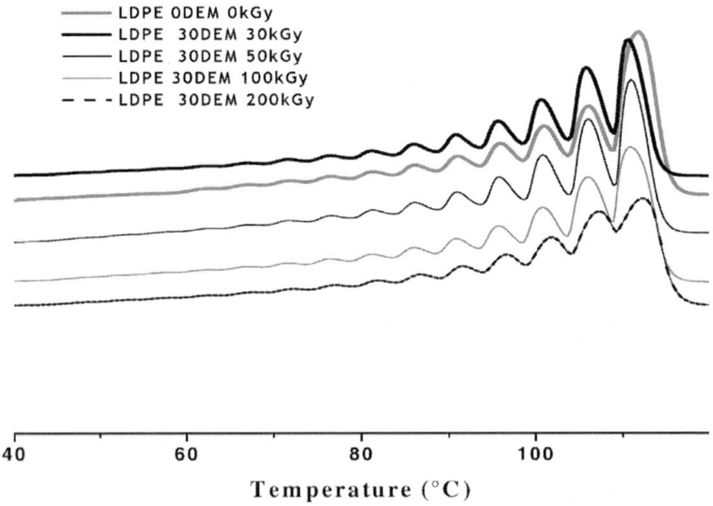

Figure 6.
SSA curves of grafted LDPE at different radiation doses, using 30% DEM in the solution.

Table 3.
Partial areas and melting peak temperatures of grafted LDPE at different radiation doses, using 30% DEM in the solution as a function of self-seeding temperature (T_s) and radiation dose.

T_s	Radiation Dose (kGy)											
	% Area						Melting Temperature (°C)					
	0	30	50	100	200	400	0	30	50	100	200	400
112	32.2	25.8	22.5	26.4	25.6	20.0	112	111	111	111	112	112
107	15.0	21.0	18.3	19.1	19.3	22.3	106	106	106	106	107	107
102	12.5	13.9	13.2	13.4	12.2	15.2	101	101	101	101	102	102
97	9.7	10.4	9.8	9.8	10.0	11.6	96	96	96	96	102	97
92	7.2	7.1	8.6	7.8	5.8	8.9	91	91	91	91	91	92
87	5.9	6.2	6.3	6.0	5.8	6.1	86	86	86	86	87	87
82	4.8	5.2	4.5	5.0	5.4	4.7	82	81	81	81	82	83
77	3.9	3.5	4.8	4.2	4.2	3.8	77	76	77	76	77	78
72	3.4	2.7	3.2	2.5	3.5	2.9	72	72	72	72	73	73

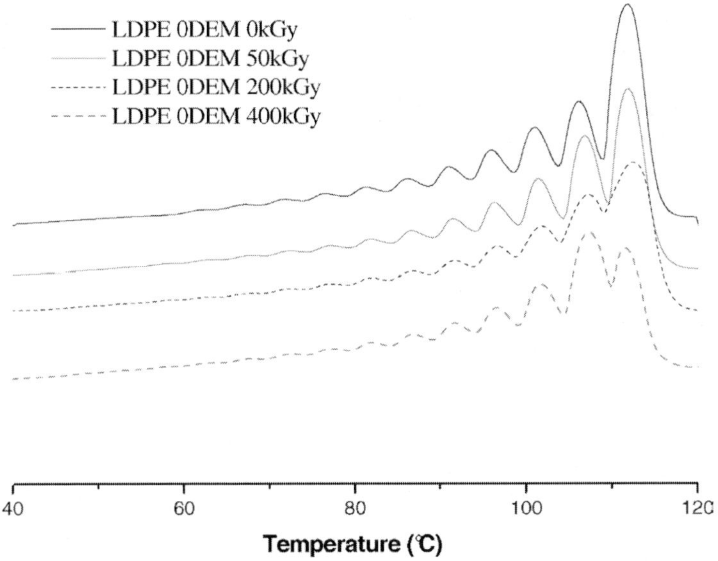

Figure 7.
SSA curves of neat LDPE irradiated at different absorbed doses.

the radiation dose and induced a significant increase in average molecular-weights. The insertion of DEM produced an interruption of the more linear sequences, which means that the functional monomer follows an insertion mechanism onto the secondary carbons.

Acknowledgements: The authors acknowledge FONACIT for its financial support through Grant F-2000001365.

[1] D. R. Paul, C. B. Bucknall, *Polymers Blends*, Eds., John- Wiley Sons, New York Vol. *1 and 2*, **2000**.
[2] N. G. Gaylord, R. Mehta, V. Kumar, M. Tazy, *J. Appl. Polym. Sci.* **1992**, *44*, 1941–1949.
[3] N. Liu, W. Baker, K. Russell, *J. Appl. Polym. Sci.* **1990**, *41*, 2285–2300.
[4] M. Aglietto, R. Bertani, G. Ruggeri, L. Segre, *Macromolecules* **1990**, *23*, 1928–1933.
[5] M. Aglietto, R. Bertani, G. Ruggeri, F. Ciardelli, *Makromol. Chem.* **1992**, *193*, 179–186.
[6] C. Rosales, R. Perera, M. Ichazo, J. Gonzalez, H. Rojas, Y. Sanchez, A. Diaz-Barrios, *J. Appl. Polym. Sci.* **1998**, *70*, 234.

[7] A. y. Simmons, W. Baker, *J. Polym. Eng. Sci* **1989**, 29, 16, 1117–1123.
[8] R. Greco, G. Maglio, P. Musto, *J. Appl. Polym. Sci.* **1989**, 37, 777–788.
[9] K. y. Ganzelved, L. Janssen, *Poly. Eng. Sci.* **1992**, 32, 7, 467–474.
[10] C. Rosales, L. Márquez, J. González, R. Perera, B. Rojas, *Poly. Eng. Sci.* **1996**, 36, 17, 2247–2252.
[11] B. Rojas, J. Fatou, C. y. Martinez, O. Laguna, *Eur. Polym. J.* **1997**, 33, 725–728.
[12] A. Chapiro, *Radiation Chemistry of Polymeric Systems*, Intercience Publishers. Londres **1962**.
[13] M. Dole, *The radiation chemistry of macromolecules*, Academic Press, **1960**.
[14] A. Charlesby, Atomic Radiation and Polymers, Pergamon, 249 Press, New York **1960**.
[15] E. Bucio, G. Cedillo, G. Burillo, T. y Ogawa, *Polym. Bull.* **2001**, 46, 115–121.
[16] B. Gupta, N. Anjum, *J. Appl. Polym. Sci.* **2001**, 82, 2629–2635.
[17] Y. Miwa, K. Yamamoto, M. Sakaguchi, S. Shimada, *Macromolecules* **2001**, 34, 2089–2094.
[18] S. Kumar, M. Pandya, *J. Appl. Polym. Sci.* **1997**, 64, 823–829.
[19] H. M. Abou Zeid, Z. I. Ali y col., *J. Appl. Polym. Sci.* **2000**, 75, 179.
[20] Y. Sánchez, C. Albano, A. Karam,, et al. *Nucl. Instr. and Meth. in Phys. Res. B.* (NIM B), **2005**, 236, 343–347.
[21] A. J. Muller, Z. H. Hernandez, M. L. Arnal, J. J. Sánchez, *Polym Bull.* **1997**, 39, 465.
[22] M. Lachtermacher, A. Rudín, *J. Appl. Polym. Sci.* **1995**, 58, 2433.
[23] Z. S. Fodor, M. Iring, F. Tüdös, T. Kelen, *J Polym Sci Polym Chem Ed*, **1984**, 22, 2530–2550.
[24] L. Márquez, I. Rivero, A. Müller, *Macromol. Chem. and Phys.* **1999**, 200, 330–337.
[25] B. Rojas de Gascue, B. Méndez, J. L. Maonosalva, J. López, V. Ruíz Sánta Quiteria, A. J. Muller, *Polymer*, **2002**, 43, 2151–2159.

Thermal Stability Evaluation of PA6/LLDPE/SEBS-g-DEM Blends

Luis Cataño,[1] *Carmen Albano,*[*1,2] *Arquímedes Karam,*[1] *Rosestela Perera,*[3] *Pedro Silva*[4]

Summary: The thermal stability of a polyamide-6/low linear density polyethylene blend (PA6/LLDPE) was studied using thermal analysis techniques. The thermogravimetric studies carried out showed that when a diethyl maleate grafted styrene-ethylene/butadiene-styrene terpolymer (SEBS-g-DEM) is added to the PA6/LLDPE blend there is an actual enhancement of the thermal stability due to the increase in the interfacial area within the blend. The Invariant Kinetic Parameter method (IKP) proved to be a qualitative technique unfolding the type of degradation mechanisms taking place in the material vicinity. Nucleation and phase boundary reactions are the kinetic models of thermal decomposition with the most significant probability of occurring.

Keywords: activation energy; compatibility; IKP; polyethylene (PE)

Introduction

Linear low-density polyethylene (LLDPE) has a widespread range of applications due to its properties arising from the comonomer addition.[1,2,3] From toy design to complex engineering applications, this polymer shows an exceptional performance. However, some of its properties limit its expansion towards new application fields, such as those requiring adhesion characteristics. Hence, there has been recent works related to the modification of the properties of LLDPE involving addition of different polymeric components.

On the other hand, PA6 is one of the engineering plastics most used on industrial applications. It has been blended with LLDPE in order to get the best of both polymers' characteristics in one material. However, because of their different natures, the PA6/LLDPE interface must be optimally increased by means of an interfacial or compatibilizing agent, such as a SEBS terpolymer.[4]

In the present work, the thermal stability of a PA6/LLDPE blend when SEBS-g-DEM is used as a compatibilizer agent was studied.

[1] Laboratorio de Polímeros, Centro de Química, Instituto Venezolano de Investigaciones Científicas, Km. 11 Carretera Panamericana, Altos de Pipe, Venezuela
E-mail: calbano@ivic.ve

[2] Universidad Central de Venezuela, Facultad de Ingeniería, Escuela de Ingeniería Química, Los Chaguaramos, Caracas, Venezuela

[3] Departamento de Mecánica, Universidad Simón Bolívar, Valle de Sartenejas, Caracas, Venezuela

[4] Laboratorio de Física de la Materia Condensada, Centro de Física, Instituto Venezolano de Investigaciones Científicas

Experimental Part

Materials

A LLDPE with 1-butene as comonomer (SCLAIR® 11D1) provided by Dupont, a PA6 (Sniamid® ADS 50), a SEBS terpolymer (Kraton®), DEM supplied by Sigma Chemical, dicumyl peroxide provided by Aldrich Chemical, Irganox B1171 and Irganox 1098 supplied by CIBA, were used.

Blend Preparation

The SEBS terpolymer was functionalized through reactive extrusion in a Berstorff ECS-2E25 corotating twin-screw extruder, using DEM as a comonomer and DCP as

Table 1.
Conditions used in blending and SEBS grafting.

Condition	Grafting	Mixing
Temperature profile (°C)	238-227-238-238-238	130-225-238-218-216
Melt Temperature (°C)	244	232
Die Temperature (°C)	225	225
Pressure drop (psi)	1000	750
Screw rate (rpm)	50	110

initiator under the conditions shown in Table 1. The grafting degree achieved in SEBS-g-DEM functionalization lies around 0.4 wt %. This grafting degree was determined according to Rosales et al.[5] by FTIR and a calibration curve.

The PA6/LLDPE (20/80% wt., respectively) and PA6/LLDPE/SEBS-g-DEM blends were prepared in a Werner & Pfleiderer ZSK-30 corotating twin-screw extruder under the conditions shown in Table 1. The compatibilizer (SEBS-g-DEM) was added at a concentration of 5% wt. Irganox B-1171 and Irganox 1098 were added before melt blending (0.5 and 0.25% wt., respectively) to avoid thermal degradation during processing.

Thermogravimetric Analysis (TGA)

Thermograms were obtained by thermogravimetric analysis on a Mettler-Toledo TGA/STDA851^e thermal analyzer under the following conditions: samples of 10 mg each were heated up to a temperature of 800 K under nitrogen atmosphere and various heating rates ($\beta_i = 5, 15$ and $20\,^\circ$C/min).

Subsequently, thermal kinetic parameters were determined by means of the Invariant Kinetic Parameters method.[6]

Molau Test

The Molau test[7] was performed on the blends in order to determine if there was a possible compatibilizing effect. Samples of the blends were dissolved in concentrated formic acid (2% w/v solution) and allowed to rest for 8 h at room temperature.

Scanning Electron Microscopy

Scanning Electron Microscopy (SEM) analyses were carried out on the cryogenically-fractured surface of compression-molded samples using a Philips 505 microscope. Samples were metallized with gold-palladium.

Invariant Kinetic Parameters (IKP) Method[8,9,10]

To determine the invariant activation energy (E_{inv}) and the pre-exponential factor (A_{inv}) the rate expression, dα/dt, was assumed to be equal to the following equation:

$$\frac{d\alpha}{dt} = k \times f(\alpha) \quad (1)$$

where α is the degree of conversion and $k = A \times \exp(-E/RT)$ according to Arrhenius law.

Eighteen apparent activation energies (E_{iv}) and pre-exponential factors (A_{iv}) were determined employing the Coats-Redfern method.[11] The IKP method is based on the principle of the compensation effect quite well reviewed in the literature. For each function $f_j(\alpha)$ proposed by the method, $log(A_j)$ versus E_j is plotted. If a compensation effect is observed, then a linear relationship is seen for each heating rate β_v, which is defined by the following expression:

$$\log A_{iv} = B_v + l_v E_{jv} \quad (2)$$

where A_{iv} and E_{jv} are the apparent pre-exponential factor and the activation energy, respectively, calculated using a function $f_j(\alpha)$ at β_v.

Inappropriate assigning of the kinetic model function results in the distortion of the kinetic parameters and in the false or superficial classification of the compensation effect.

The values B_v and l_v are calculated from the intercept and the slope of the straight lines obtained by Equation (2). Lesnikovich

Table 2.
Degradation functions used in the IKP method.[6]

Kinetic models	$f_j(\alpha)$	$g_j(\alpha)$	Observations
Nucleation and nucleus growth	$\frac{1}{n}(1-\alpha)(-\ln(1-\alpha))^{1-n}$	$(-\ln(1-\alpha))^n$	S1-$n=1/4$
			S2-$n=1/3$
			S3-$n=1/2$
			S4-$n=2/3$
Phase boundary reaction	$(1-\alpha)^n$	$1-(1-\alpha)$	S6 Plane Symmetry
		$2\left[1-(1-\alpha)^{1/2}\right]$	S7 Cylindrical Symmetry
		$3\left[1-(1-\alpha)^{1/3}\right]$	S8 Spherical Symmetry
Diffusion	$1/2\alpha^{-1}$	α^2	S9 Plane Symmetry
	$(-\ln(1-\alpha))^{-1}$	$(1-\alpha)\ln(1-\alpha)+\alpha$	S10 Cylindrical Symmetry
	$3/2\left[(1-\alpha)^{-1/3}-1\right]^{-1}$	$1-2/3\alpha-(1-\alpha)^{2/3}$	S11 Spherical Symmetry
	$3/2(1-\alpha)^{1/3}\left[(1-\alpha)^{-1/3}-1\right]^{-1}$	$\left[(1-\alpha)^{1/3}-1\right]^2$	S18 Jander's Type
Potential law	$(1/n)\alpha^{1-n}$	$\alpha^n(0<n<2)$	S12-$n=1/4$
			S13-$n=1/3$
			S14-$n=1/2$
			S17-$n=3/2$
Reaction order	$(1-\alpha)$	$-\ln(1-\alpha)$	S5-$n=1$
	$(1/n)(1-\alpha)^{1-n}$	$1-(1-\alpha)^{1/2}$	S15-$n=1/2$
		$1-(1-\alpha)^{1/3}$	S16-$n=1/3$

and Levchik[8,9] discussed the significance of these values and demonstrated the following relationships:

$$B_v = \log(k_v) \tag{3}$$

$$I_v = (2.3RT_v)^{-1} \tag{4}$$

where k_v is the rate constant at of the system at the temperature T_v; these two parameters are characteristic of the experimental conditions.

The curves $log(k_v)$ versus $1/T_v$ are plotted in order to calculate the intercept and slope of this equation:

$$\log(k_v) = \log(A_{inv}) - E_{inv}/2.3RT_v \tag{5}$$

which finally results in the values of the invariant activation energy and the pre-exponential factor of the evaluated sample.

The probabilities associated with the 18 degradation functions proposed in the literature are presented in Table 2. The degradations are complex phenomena and must be represented by a set of functions instead of a single one.

Results and Discussion

The thermograms of PA6/LLDPE/SEBS-g-DEM are presented in Figure 1 for three heating rates under nitrogen atmosphere. Thermal decomposition proceeded in a single step; however, for the higher heating rate, a slight change in the slope as the sample heats up can be seen. The main decomposition step takes place in a broad temperature range (680-780 K). Moreover, the TGA thermograms of PA6/LLDPE/SEBS-g-DEM shift towards the right as the heating rate increases in the samples. No significant difference could be noticed between the thermograms with and without the interfacial agent.

DTG curves of PA6/LLDPE/SEBS-g-DEM are shown in Figure 2. Degradation rates are being shifted to higher temperatures due to the use of higher heating rates which promote a difference in the temperature profile in the sample. As the heating rate increases, there are some differences in the occurrence of the degradation reaction mechanisms. For instance, the DTG peak shifts towards the right and decrease its degradation rate as the heating rate increases (Figure 2(a)). On the other hand, PA6/LLDPE DTG curves (Figure 2(b)) do not exhibit a remarkable change on the degradation rate, possibly due to the fact that there is no interfacial area enough to reach

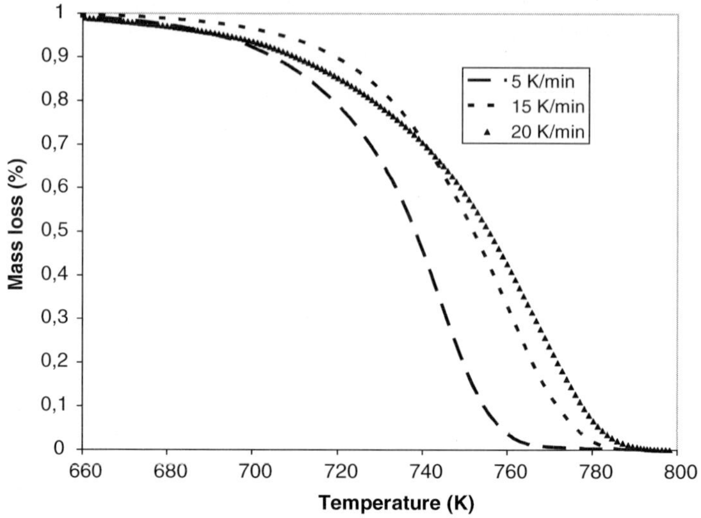

(a)

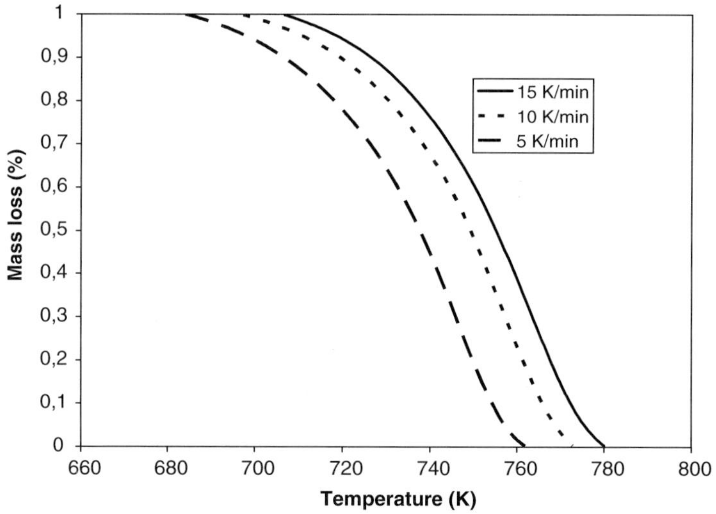

(b)

Figure 1.
(a) PA6/LLDPE/SEBS-g-DEM and (b) PA6/LLDPE thermograms.

an activation of different decomposition models.

This phenomenon is in agreement with those reported in the literature[12] on the basis of using several heating rates with the purpose of calculating the thermal kinetic parameters.

The IKP method uses the compensation effect in order to determine the values that would serve us to calculate the invariant parameters independent of the heating rate employed. Figure 3 illustrates the linear relationship found between $log(A_j)$ and E_j for PA6/LLDPE/SEBS-g-DEM blend at

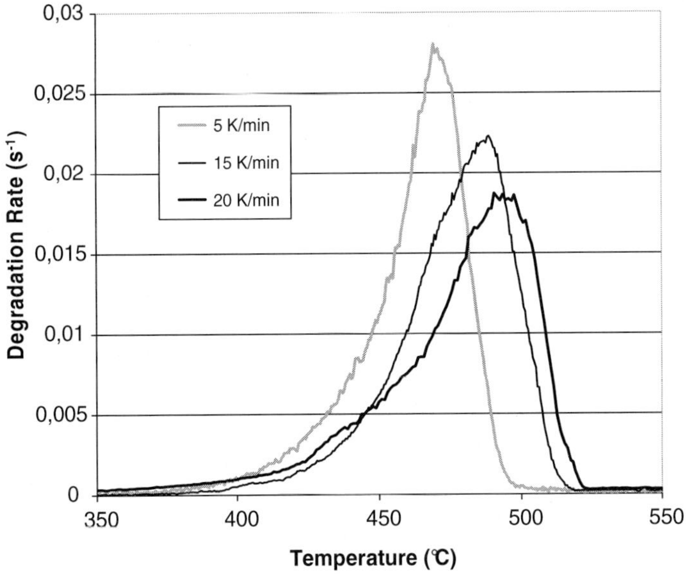

(a)

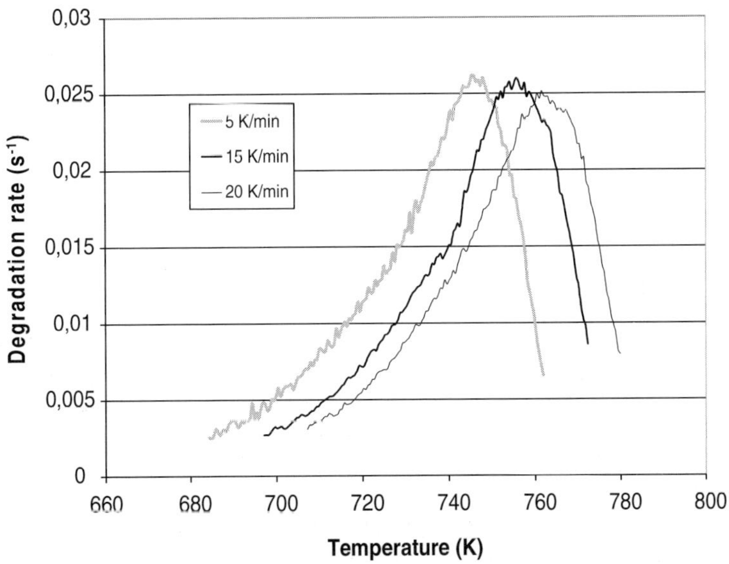

(b)

Figure 2.
DTG curves of (a) PA6/LLDPE/SEBS-g-DEM and (b) PA6/LLDPE.

several heating rates. A similar trend is observed for PA6/LLDPE blend.

Once the straight line is obtained from the apparent kinetic parameters determined through the Coats-Redfern method, the values B_v and I_v are then calculated in order to plot a new straight line whose slope and intercept correspond to the invariant activation energy and pre-exponential factor, respectively.

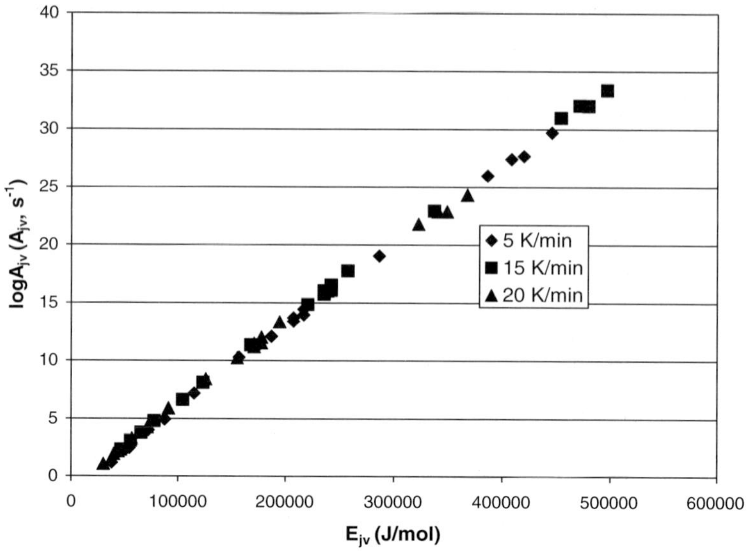

Figure 3.
Compensation effect observed in PA6/LLDPE/SEBS-g-DEM.

The invariant kinetic parameters obtained are depicted in Table 3. These results show that the addition of the interfacial agent in the PA6/LLDPE blend improves the thermal stability of the material.

The calculated invariant activation energy for the PA6/LLDPE/SEBS-g-DEM was 540 KJ/mol, which shows an increase of nearly 100% when compared to that of the PA6/LLDPE blend. This could be due to the better phase dispersion achieved when the SEBS-g-DEM is added to the blend, which is somehow improving the thermal stability of the material. Additionally, the IKP method allowed determining the probabilities of several degradation mechanisms to occur. The kinetic models assumed in the thermal decomposition are shown in Table 2.

Degradation profiles are governed by a combination of nucleation and phase boundary reactions in both of the studied blends. The probabilities of phase boundary reaction and nucleation mechanisms (Figure 4 and 5) were raised by 26%, accounting for an increase of the surface area between PA6 and LLDPE on the samples, evidencing that SEBS-g-DEM could effectively be acting as a compatibilizing agent in this blend, supporting the increase of the Ea. The Molau test also evidenced this phenomenon. Differences on material reactivities cause the blend to exhibit a thermal resistance.[13] Nucleation and nucleus growing reaction mechanisms were the most probable to take place in both samples, with and without SEBS-g-DEM.

Table 3.
Invariant kinetic parameters obtained by the IKP method.

Sample	E_{inv} (KJ/mol)	log A_{inv}
PA6/LLDPE	283	19.03
PA6/LLDPE/SEBS-g-DEM	519	35.15

On the other hand, Figure 6 presents the dependence of the kinetic functions taking place in the degradation with the degree of conversion achieved by the material. The equation plotted in Figure 6 is given by:

$$f(\alpha) = \sum_{j=1to18} (\%) f_j(\alpha) \tag{6}$$

The continuous line represents the behavior of the PA6/LLDPE, while the

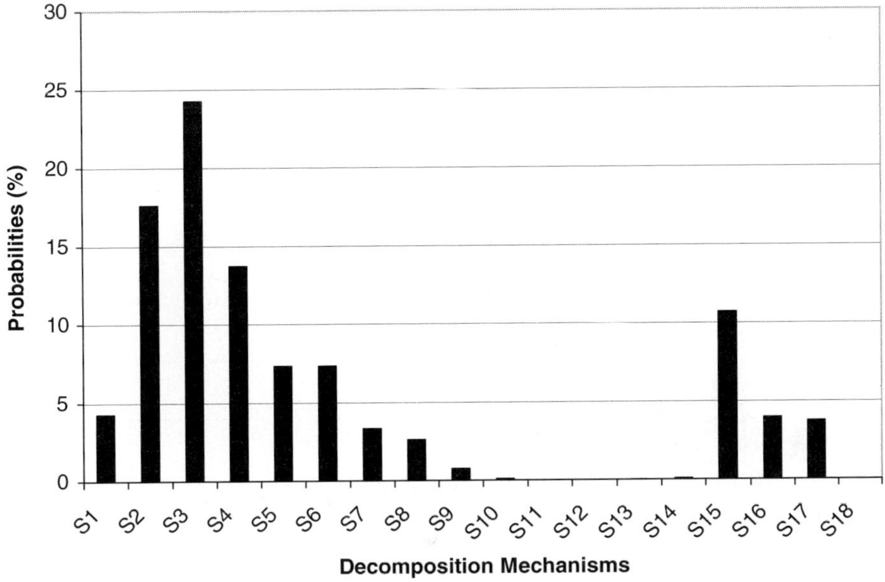

Figure 4.
Probability distribution of the 18 kinetic models used in the IKP method for the PA6/LLDPE blend.

dotted curve belongs to the PA6/LLDPE/SEBS-g-DEM. The influence of the addition of SEBS-g-DEM turns significant at early degradation stages when it starts decomposing in its surface. However, this influence decreases as the degree of conversion increases. On the other hand, for the PA6/LLDPE blends, the curve stays flat indicating that there is not a significant dependence of those kinetic models

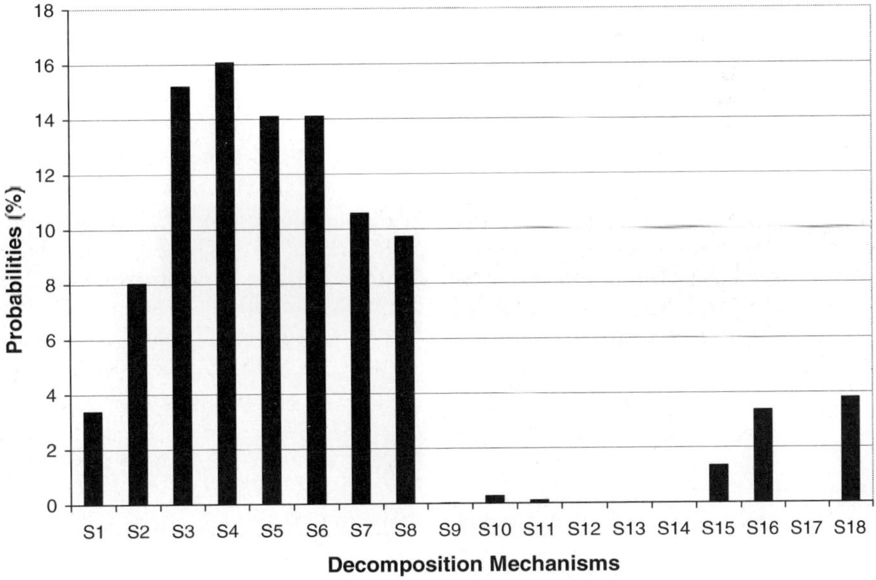

Figure 5.
Probability distribution of the 18 kinetic models used in the IKP method for the PA6/LLDPE/SEBS-g-DEM blend.

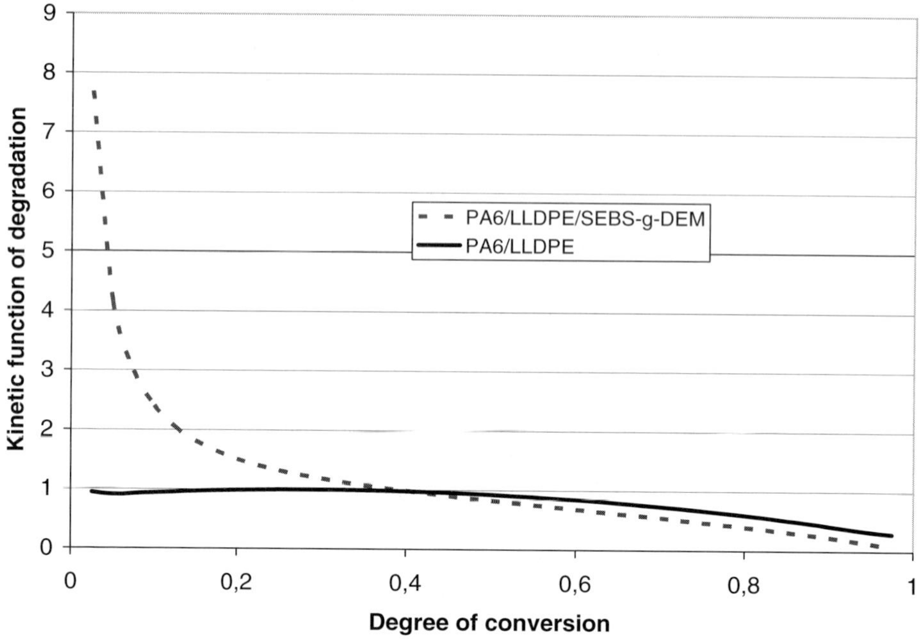

Figure 6.
Dependence of the f (α) with the degree of conversion (α).

occurring with the degree of conversion. Still, at lower conversion degrees, the PA6/LLDPE/SEBS-g-DEM exhibits higher function values, but when the conversion degree rises ($\alpha > 0.4$), the function values of the PA6/LLDPE/SEBS-g-DEM decreases even to lower values than those of the PA6/LLDPE, which is attributed to a protective barrier[14] due to the better phase dispersion achieved by the interfacial agent. This statement is confirmed by scanning electron microscopy (SEM). SEM micrographs are shown in Figure 7.

The degradation rates V versus α and T:

$$V = A_{inv} \times \exp(-E_{inv}/RT) \times \sum_{j=1to18} (\%) f_j(\alpha) \quad (7)$$

are plotted in Figure 8 and 9 for PA6/LLDPE and PA6/LLDPE/SEBS-g-DEM

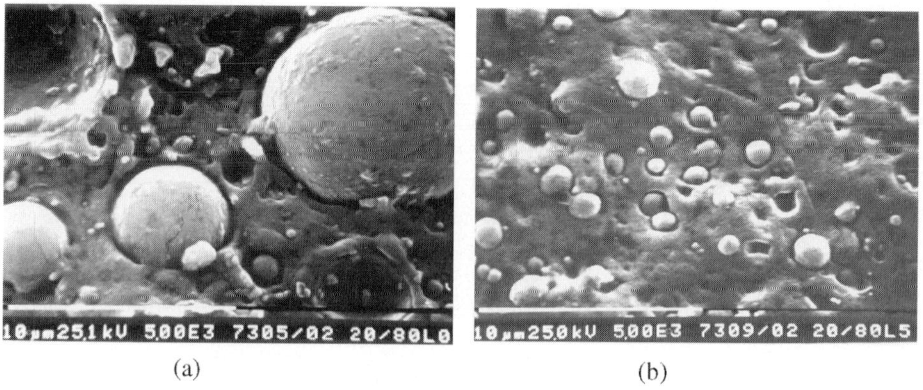

Figure 7.
SEM micrographs of (a) PA6/LLDPE and (b) PA6/LLDPE/SEBS-g-DEM.

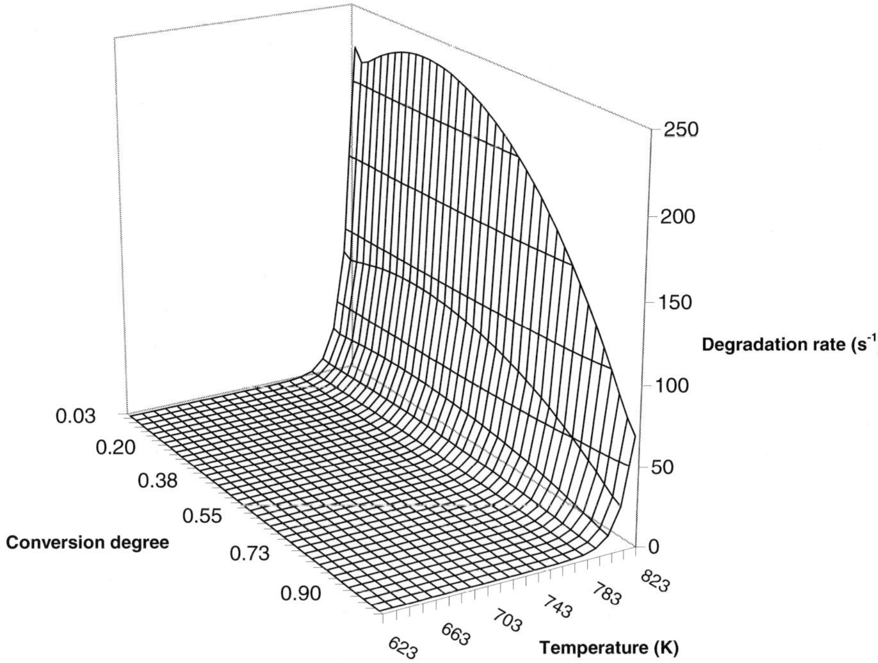

Figure 8.
Degradation rate of PA6/LLDPE versus conversion degree and temperature.

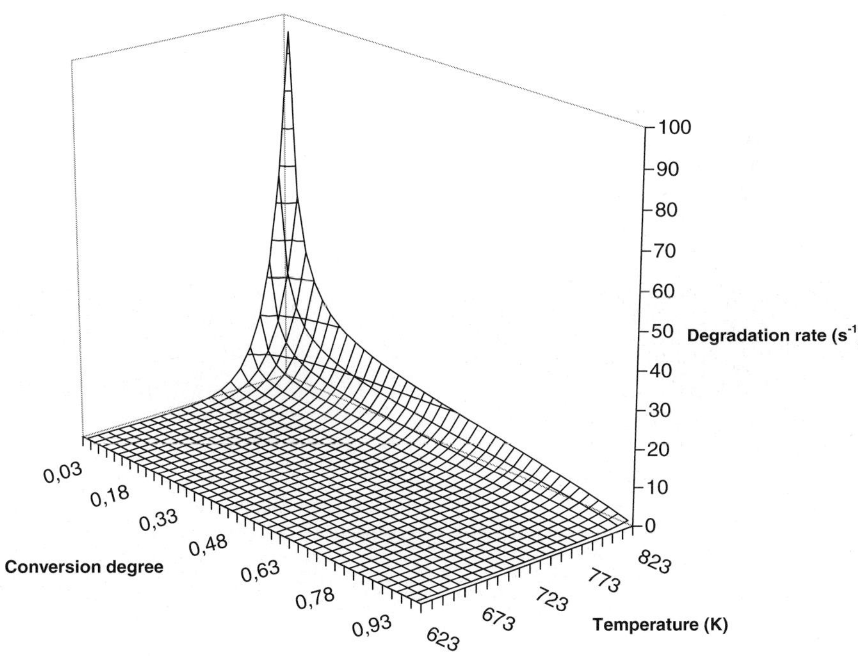

Figure 9.
Degradation rate of PA6/LLDPE/SEBS-g-DEM versus conversion degree and temperature.

samples, respectively. The 3D plots show that the blend without the interfacial agent decomposes faster at higher temperatures and at any given conversion degree than the PA6/LLDPE/SEBS-g-DEM blend. Degradation rates are lower in the blend with the interfacial agent, evidencing that its inclusion enhances the thermal stability of the blend.

Conclusions

The thermogravimetric analysis carried out in these samples showed that when the SEBS-g-DEM is added to the PA6/LLDPE there is an actual enhancement of the thermal stability due to the increase in the interfacial area within the blend. The IKP method proved to be a qualitative technique evidencing the type of degradation mechanisms taking place in the material vicinity. Nucleation and phase boundary reactions are the kinetic models on the thermal decomposition more likely of occurring. Statistical calculations along with the Molau test evidenced that the inclusion of SEBS-g-DEM in the PA6/LLDPE increases the thermal stability.

Appendix[15]

The degradation is modeled by computing the probabilities associated with the 18 degradation functions showed in Table 2. Degradation of a polymer material often cannot be represented with a single degradation function. The kinetic functions $f_j(\alpha)$ may then be discriminated using the $\log A_{inv}$ and E_{inv} values obtained. Having n of the ith of the experimental values of $(d\alpha/dT)_{iv}$, the residual sum of squares for each $f_j(\alpha)$ and for each heating rate β_v may be computed as:

$$(n-1)Sj^2_{jv} = \sum_{i=1}^{i=n} \left| \left(\frac{d\alpha}{dT}\right)_{iv} - \frac{A_{inv}}{\beta_v} \exp\left(-\frac{E_{inv}}{RT_{iv}}\right) f_j(\alpha_{iv}) \right|^2 \quad \text{(A1)}$$

The most probable function is then chosen by the average minimum value of $\overline{S}_j$ defined by the relationship,

$$\overline{S}_j = \frac{1}{p} \sum_{v=1}^{v=p} S_{jv} \quad \text{(A2)}$$

where p is the number of heating rates used. The probability associated with each value $f_j(\alpha)$ can be calculated by defining the ratio,

$$F_j = \frac{\overline{S}_j^2}{\overline{S}_{\min}^2} \quad \text{(A3)}$$

where $\overline{S}_{\min}^2$ is the average minimum of residual dispersion. This ratio obeys the F distribution,

$$q(F_j) = \frac{\Gamma(v)}{\Gamma^2(v/2)} \frac{F_j^{(v/2)-1}}{(1+F_j)^v} \quad \text{(A4)}$$

where n is the number of degrees of freedom equal for every dispersion and Γ is the gamma function. It is interesting to note that the average of the residual dispersion, and not simply the residual dispersion, was chosen to define the ratio F_j because the average $\overline{S}_j^2$ is a good non-biased estimate of all S_{jv}^2 values and gives a better statistic representation of the process.

The probabilities of the jth function are computed on the assumption that the experimental data with L kinetic functions are described by a complete and independent system of events:

$$\sum_{j=1}^{j=L} P_j = 1 \quad \text{(A5)}$$

Therefore we obtain:

$$P_j = \frac{q(F_j)}{\sum_{j=1}^{j=L} q(F_j)} \quad \text{(A6)}$$

[1] C. Albano, J. Trujillo, A. Caballero, O. Brito, *Polymer Bulletin* **2001**, 45(6), 531.
[2] C. J. R. Verbeek, *Material Letters* **2002**, 52, 453.
[3] S. Bhadrakumari, P. Predeep, *Supercond. Sci. Technol.* **2006**, 19, 808.
[4] J. Reyes, C. Albano, M. Claro, D. Moronta, *Radiation Physics and Chemistry* **2003**, 63, 435.
[5] C. Rosales, H. Rojas, R. Perera, A. Sánchez, *J.M.S.-Pure and Applied Chemistry* **2000**, A37(10), 1227.

[6] S. Bourbigot, R. Delobel, M. Le Bras, D. J. Normand, *Chim. Phys.* **1993**, *90*, 1909.
[7] G. E. Molau, *J Polym. Sci.* **1965**, 3(4), 1267.
[8] A. J. Lesnikovich, S. V. Levchik, *J. Therm. Anal.* **1983**, *27*, 83.
[9] A. J. Lesnikovich, S. V. Levchik, *J. Therm. Anal.* **1985**, *30*, 667.
[10] S. V. Levchik, G. F. Levchik, A. J. Lesnikovich, *Thermochim Acta* **1985**, *77*, 157.
[11] A. W. Coats, J. P. Redfern, *Nature* **1964**, *201*, 68.
[12] R. Serra, J. Sempere, R. Nomen, *Thermochim Acta* **1998**, *316*, 37.
[13] V. V. Boldyrev, M. Bulens, B. Delmon, The control of the reactivity of solids. Studies in Surface Science and Catalysis 2. Elsevier Scientific Publishing Company **1979**.
[14] F. Dabrowsky, S. Bourbigot, R. Delobel, M. Le Bras, *European Polymer Journal* **2000**, *36*, 273.
[15] S. Bourbigot, X. Flambard, S. Duquesne, *Polymer International* **2001**, *50*, 157.

 DOI: 10.1002/masy.200751115

Using Solvents to Improve the Chemical Shift Differences Between Short-Chain Branch Methines and Long-Chain Branch Methines in Polyethylene Copolymers

Dan Baugh,[1] O. David Redwine,[2] Angela Taha,[1] Ken Reichek,[1] Janece Potter[1]*

Summary: Detection and quantification of long-chain branches in some polyethylene copolymers is challenging due to the near coincidence of the chemical shifts for the carbons at the short-chain and long-chain branches present in these copolymers. The small chemical shift difference can be enhanced by changes in solvent and temperature. This allows one to use lower field magnets for some copolymers. Results are presented comparing several solvents and blends at a variety of temperatures using 500, 600 and 750 MHz spectrometers.

Keywords: branched; LLDPE; NMR; polyethylene (PE); solvent

Introduction

Ethylene-octene and ethylene-hexene copolymers are common linear low density polyethylene (LLDPE) polymers, representing over 75% of the total LLDPE market. As new materials are developed and commercialized for this growing market (6% global annual growth rate), it is expected that ethylene-octene and ethylene-hexene copolymers will continue to constitute a major portion of it. To fully evaluate these new materials, it is of critical value to expand and improve the available analytical methods for long-chain branching (LCB) analysis. Detection and quantification of LCB in these copolymers is challenging due to the near coincidence of the chemical shifts for the carbons at the short-chain branches (SCB) and long-chain branches present in these copolymers (Figure 1).

Solvent screening experiments have shown that solvent and temperature effects on the shift difference between the short and long-chain branch methines are quite significant. The small chemical shift difference observed in samples prepared using high boiling chlorinated aromatic solvents can be enhanced by changes in solvent and temperature. This allows one to use lower field magnets for some copolymers. A combination of resolution enhancement, solvent selection, sample temperature and high magnetic field (188 MHz, 150 MHz and 125 MHz ^{13}C) have been used to achieve enhanced resolution for the respective methine carbons of these two branch types. Quantification of LCB was validated by the measurement of a controlled sample that contained a known amount of long-chain branches.

This work is a continuation of the extensive application of NMR, rheology, and solution property methods to characterize polyolefins.[1] This capability will allow extended characterization of competitive copolymers and enhanced materials science understanding of new polyolefin materials.

[1] The Dow Chemical Company, 2301 N. Brazosport Blvd., Freeport, Texas 77541
E-mail: DWBaughIII@dow.com
[2] The Dow Chemical Company, 1897 Bldg., Midland, Michigan 48667

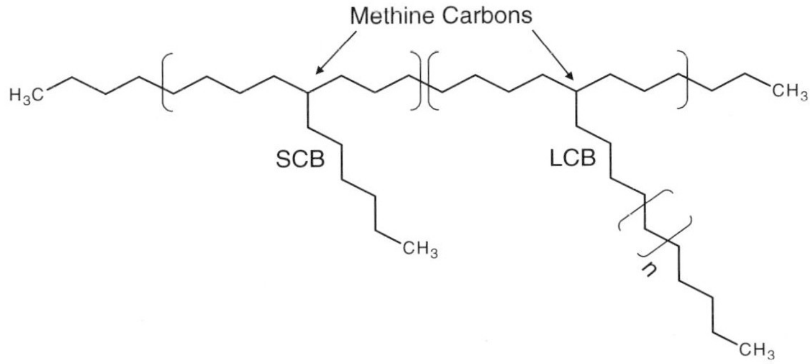

Figure 1.
Structure illustrating SCB and LCB methine carbons in an ethylene-octene (EO) copolymer.

Experimental Part

Sample Preparation

A total of 0.45 g of polymer is weighed into a 10 mm o.d. NMR tube. The solvent is added to the NMR tube. The tube is capped and placed in a heating block. The nominal temperature of the heating block is 150 °C. The hot sample is frequently mixed using a vortex mixer or a Buchi oven with rotating sample holder. The solvents used in this study are marginal polyethylene solvents; therefore preparation of a homogeneous sample is tedious. A significant amount of time and mixing is required to achieve solutions of sufficient homogeneity.

Sample Description

The total polymer weighed is a mixture of a linear low density ethylene-hexene or ethylene-octene copolymers and a long-chain branched homopolymer containing a known amount of branching. The various solvents used in this study are summarized in Table 1.

Data Acquisition

Nalorac high temperature 10 mm Z-spec probes were used at both 600 MHz and 750 MHz. Prior to data acquisition, the observed pulse widths are verified for both the ^{13}C and ^{1}H channels. Typical acquisition parameters for the Varian Inova 500 and 600 MHz and Varian Unity Plus 750 MHz NMR spectrometers are 64 K data file, 3.3 sec repetition rate and a 90° pulse width.

Data Processing

Chemical shift assignments are based on previously determined values of 30.0 ppm for the backbone methylene signal.[2,3] All data processing is done using NUTS (NMR Utility Transform Software) available from Acorn NMR.[4] The data files were processed with a weak Gaussian apodization function by setting LB = −0.7 Hz and GF = 0.10. This effectively changed the lineshape from Lorentzian to Gaussian without significant resolution enhancement except at the peak base.

Results and Discussion

The primary feature of interest when quantifying LCB in these copolymers is the separation between the LCB CH and the CH of the SCB associated with the EOE

Table 1.
Solvent compositions used in this study.

Name	Composition
TCE/ODCB	50/50 by weight 1,1,2,2-tetrachloroethane (TCE-d_2) and *o*-dichlorobenzene (ODCB).
Biphenyl	90/10 wt/wt biphenyl/biphenyl-d_{10}
Mesitylene	90/10 wt/wt 1,3,5 trimethyl benzene and 1,3,5 trimethyl benzene-d_{12}
Biphenyl/TCE	3.6/1 wt/wt biphenyl/TCE-d_2
Naphthalene	90/10 wt/wt naphthalene/naphthalene-d_8

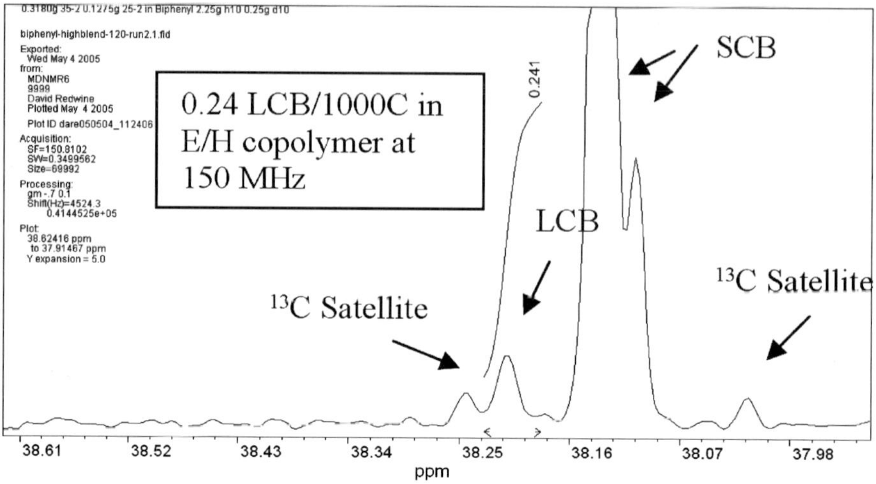

Figure 2.
A 150 MHZ ^{13}C spectrum of a sample containing a mixture of a linear ethylene-hexene copolymer and a long-chain branched HDPE. The LCB spike has a known level of 0.24 LCB per 1,000 total carbons.

or EHE triad. It is this chemical shift separation that has been adjusted by using various solvents in this work. It is also important to maintain some resolution between the LCB CH resonance and the ^{13}C satellite resonance of the SCB CH resonance. The optimum placement of the LCB methine signal is approximately 8.5 to 9 Hz down-field from the SCB CH resonance as shown in Figure 2 for an ethylene-hexene copolymer in biphenyl/TCE at 150 MHz ^{13}C. Using chlorinated aromatic solvents this degree of separation requires a magnetic field in excess of

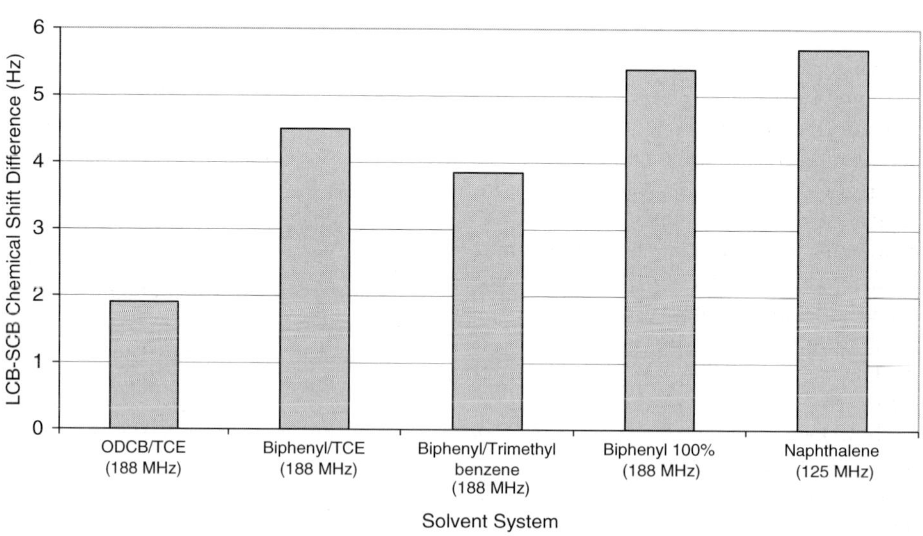

Figure 3.
A bar chart comparing the LCB-SCB methine separation at 120 °C for an ethylene-octene copolymer. The optimum solvent, naphthalene, gives a separation at 125 MHz which is over 2.5X the separation observed in the usual chlorinated solvent system at 188 MHz.

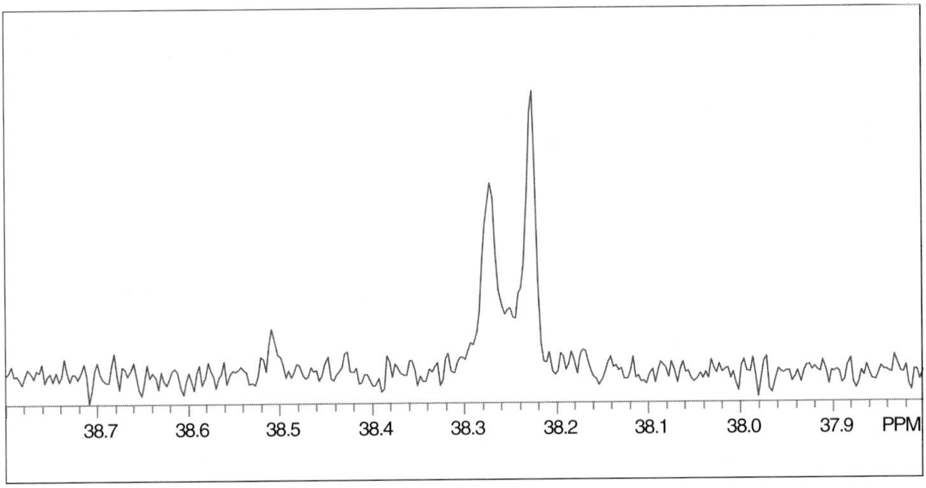

Figure 4.
A 125 MHZ ^{13}C spectrum of sample containing a mixture of a linear ethylene-octene copolymer and a long-chain branched HDPE. The LCB spike has a known level of 0.24 LCB per 1,000 total carbons. This spectrum was obtained in the optimum solvent, naphthalene, and shows a separation of 5.7 Hz.

750 MHz. A comparison of the LCB-SCB methine resonance separation is plotted in Figure 3. Figure 4 shows the resolved SCB and LCB methine signals for an EO copolymer.

Conclusions

The NMR methods described in this report produce ^{13}C spectra that satisfy the resolution and sensitivity requirements for LCB analysis in ethylene-hexene or ethylene-octene copolymers produced by transition metal polymerization. Dramatic shift enhancements have been observed for a variety of solvents using biphenyl and naphthalene as the primary solvents. Limited understanding of this effect is based on a rotational isomeric state conceptual model of the CH chemical shifts. The observed shifts for the SCB methine and the LCB methine are both average values over the population of the various rotomers possible for the main-chain and side-chain carbons near the methine carbon. Biphenyl and naphthalene produce a different population distribution and therefore result in different shifts compared to TCE/ODCB. It is believed that the differences occur mainly at the ends of the side-chains. These carbons are close enough to the side-chain end to give some selectivity in solvation for the SCB side-chain. This primary difference causes a secondary effect of changed steric effects at the CH carbon resulting in changes to the SCB and LCB methine resonance frequency.

Regardless of the mechanism at play we have succeeded in developing solvent systems which give the optimum separation of approximately 8.5–9 Hz for ethylene-hexene copolymers at 150 MHz ^{13}C (biphenyl/TCE), and ethylene-octene at 188 MHz ^{13}C (naphthalene).

Acknowledgements: The authors also wish to acknowledge the assistance of Drs. Shanmin Zhang and David Gorenstein at the University of Texas Medical Branch, Galveston.

[1] P. M. Wood-Adams, J. M. Dealy, A. W. deGroot, O. D. Redwine, *Macromolecules* **2000**, 33, 7489.
[2] M. DePooter, G. Smith, K. K. Dohrer, K. F. Bennett, M. D. Meadows, H. P. Schouwenaars, R. A. Geerards, *J. Appl. Polym. Sci.* **1991**, 42, 399.
[3] W. Liu, D. G. Ray, P. L. Rinaldi, *Macromolecules* **1999**, 32, 3817.
[4] NMR Utility Transform Software (NUTS) 2D Professional Version – 20050824 (**2005**), Acorn NMR Inc., Livermore, CA 94550.

DOI: 10.1002/masy.200751116

The Effect of Feed Composition of Styrene and Acrylic Acid on the Properties of Modified Low Density Polyethylene

*Shenglong Ding, Mingzhu Liu**

Summary: Low density polyethylene (LDPE) was functionalized with different molar ratios of styrene (St) to acrylic acid (AA) in the presence of dicumyl peroxide (DCP) in the molten state. The resultant LDPE was characterized by gel content and torque analysis. The results showed that the gel content of polymers grafted with small molar ratios (St/AA = 0.5) was always higher than those grafted with the equimolar St/AA ratio. The effect of DCP amount and AA concentration on the grafting degree of AA was investigated. The suitable DCP amount and AA concentration was obtained. Functionalized LDPE [LDPE-*g*-(AA-St)] was characterized by Fourier transform infrared spectroscopy (FTIR), melt flow rate (MFR), water contact angle and capillary rheometry. The results showed that both MFR and grafting degree of AA of LDPE-*g*-(AA-St) was the highest when an equimolar AA/St ratio was used, but when mixtures of St and AA (St/AA = 1/10) were loaded, the water contact angle of the film prepared from the LDPE-*g*-(AA-St) was the smallest, which indicated that the hydrophicity of the film surface not only depended on the grafting degree, but also the molar ratios of the hydrophobic/hydrophilic groups.

Keywords: acrylic acid; graft copolymers; modified polyethylene; rheology; styrene

Introduction

Low density polyethylene (LDPE) is a commodity polymer used in extrusion operations such as blown film, coating, blow molding, and foaming for its low cost. However, its application is greatly limited due to its lack of polarity and lower reactivity. Chemical modification of LDPE is an important method to expand the applications of LDPE and generate value-added materials with improved mechanical, thermal, polar, and chemical properties.

Modified LDPE can react with chemical coupling agents to increase the polymer interfacial interactions with inorganic fillers and its miscibility with polar polymers as well as to produce polyolefin with improved adhesion and dyeability.

Department of Chemistry, Lanzhou University, Lanzhou 730000, P. R. China
E-mail: m-zliu@163.com

From an industrial point of view, free radical initiated functionalization of polyolefins in the molten state had received much attention over the past decades. Acrylic acid (AA),[1,2] glycidyl methacrylate (GMA),[3,4] and maleic anhydride (MAH) were employed widely as monomers to enhance polarity and reactivity of polyolefin.[5–7] Grafting copolymerization of LDPE with AA was investigated by Ghosh[8] and the conclusion was drawn that the grafting reaction in the molten state was first order with respect to AA concentration. Besides, the information of pendant structures of LDPE-*g*-AA suggested that AA monomers grafted onto the LDPE backbone were several units long[9] and that the grafted AA could act as a nucleating agent for crystallization of LDPE.[10]

Styrene (St) as a second monomer in the melt-grafting system not only increases the graft degree of most monomers on polyolefin, but also reduces crosslinking of

LDPE. However, the addition of St will decrease the hydrophobicity of the LDPE-*g*-(AA-St) surface.

Grafting AA onto LDPE has been reported in the literature, but there are few articles that study the effect of the amount of St on the hydrophilicity and rheological properties of LDPE-*g*-(AA-St). In this work, the effects of feed composition of St and AA, and concentrations of initiator and monomer on the grafting degree of AA and extent of crosslinking of modified LDPE were systematically studied. LDPE-*g*-(AA-St) would be used as compatibilizer for blends of LDPE and nylon 6.

Experimental Part

Materials

LDPE with MFR = 2 g/10 min (190 °C, 2.16 kg) was supplied by Lanzhou Petro-Chemical Co. (China). Reagent-grade acrylic acid (AA, 99% purity) and styrene (St, 99% purity) were purchased from Tianjin Institute of Chemical Reagents (China) and used without further purification. Dicumyl peroxide (DCP), purified by recrystallization from ethanol prior to use, was purchased from Shanghai Reagent Co. (China). DCP has a half-time of about 1.45 min at 175 °C. The modified LDPEs using AA, and St and AA, were designated LDPE-*g*-AA and LDPE-*g*-(AA-St), respectively. St/AA represents the molar ratio of styrene to acrylic acid.

Melt Grafting

The grafting reactions were carried out in a Haake Rhemix 600P batch mixer, equipped with roller blades and a mixing head with a volumetric capacity of 69 cm^3. A 45 g charge of the vacuum-dried LDPE at 60 °C was blended with the desired amount of St, AA and DCP, then fed into the mixer which had already been adjusted to the optimum conditions, which for processing in the Haake Rhemix 600P was selected from our previous work:[11] temperature of 170 °C, mixing speed of 80 rpm, and reaction time of 10 min. After mixing, the samples were taken from the chamber and quenched with liquid nitrogen to stop further reactions.

Purification and Characterization of the LDPE-*g*-(AA-St)

About 3.0 g of gross grafted products were dissolved in 150 mL boiling xylene and precipitated in 500 mL acetone to remove unreacted monomers and homopolymers formed during grafting. Then the polymer was extracted with acetone for 6 hours once again. All purified polymer samples were dried to a constant weigh at 50 °C under vacuum.

The Fourier transform infrared (FTIR) spectra of the purified polymers were recorded with a FTIR spectrometer (AVATAR 360 FTIR, Nicolet, US). The resolution was 4.0 cm^{-1}, and the scanned wave number ranged from 4000 to 400 cm^{-1}. The purified products were pressed into thin films at 180 °C for FTIR measurement. From the FTIR spectrum, the absorbance ratio (R_a) of the area of the bands at 1709 cm^{-1} and 1467 cm^{-1} represents the relative grafting degree of the AA. R_a was calculated with the following equation:

$$R_a = A_{1709}/A_{1467} \quad (1)$$

where A_{1709} is the peak area of absorbance at 1709 cm^{-1}, characteristic of the carbonyls from AA, and A_{1467} is that of the absorbance at 1467 cm^{-1}, characteristic of the CH_2 groups.

The melt flow rates (MFR) of the purified polymers were determined using XRZ-400 type MFR equipment at 190 °C with a load of 2.16 kg according to the ASTM 1238–89 standard.

The gel content was obtained by the following measurement: An unpurified sample (100–120 mg) was packed in a preweighed nickel net (120 mesh) and put into a Soxhlet extractor, extracted with boiling xylene for 24 hours, then dried in a vacuum oven at 60 °C until its weight was a constant. The gel content was calculated with the following equation:

$$\text{Gel content } (\%) = (W_s - W_n)/W_p \times 100\% \quad (2)$$

where W_s, W_n, and W_p represents the total weight of polymer and nickel net after being extracted, the weight of nickel net, and the weight of polymer, respectively.

LDPE film specimens were cast into thin films of about 0.1 mm thickness by dropping a dilute solution of ca. 5 wt% concentration in xylene onto clean glass slides and evaporating the solvent. The films were further dried under vacuum for a few days before use. The film specimens were carefully peeled from each glass slide just before measuring the contact angle on the glass side of the films.

The contact angles of double-distilled water on the control LDPE and grafted LDPE films were measured at ambient temperature. Liquid droplet was gently placed on the specimen. The height (h) and the base (w) of the droplet were measured from the photograph and geometric considerations. Each contact angle was the average of at least eight measurements.

The rheological behavior of LDPE and LDPE-*g*-(AA-St) was investigated with an XLY-II flow tester (capillary rheometer) (Jinlin University Science and Educational Instrument Plant, Changchun, China). The fixed-temperature method was used. The nozzle diameter was 1 mm, and the nozzle length was 40 mm. The operation temperature was fixed at 190 °C, with experimental loads of 20, 30, 40, 50, 60 and 70 kgf/cm^2. More detailed information cand be found in elsewhere.[12]

Results and Discussion

FTIR Spectroscopy Analysis of the Grafted LDPE

The FTIR spectra of LDPE, LDPE-*g*-AA, and LDPE-*g*-(AA-St) are shown in Figure 1. In the cases of LDPE-*g*-AA and LDPE-*g*-(AA-St), the new absorption band at 1709 cm^{-1} was observed, which can be assigned to the absorption of the carbonyl groups (–C=O) of AA. This new peak confirmed that the AA monomer was grafted onto the LDPE chain. The absorption bands at 1369 cm^{-1} and 1467 cm^{-1}, were also observed, which can be attributed to scissor vibrations of methyl (–CH_3) and methylene (–CH_2) groups. The absorption band at 719 cm^{-1} corresponds to the swing vibration of methylene groups. In the case of LDPE-*g*-(AA-St), the characteristic absorption band of benzene ring groups at 700 cm^{-1} overlapped that of 719 cm^{-1}.

The intensity of the carbonyl absorption band at 1709 cm^{-1} for LDPE-*g*-(AA-St)

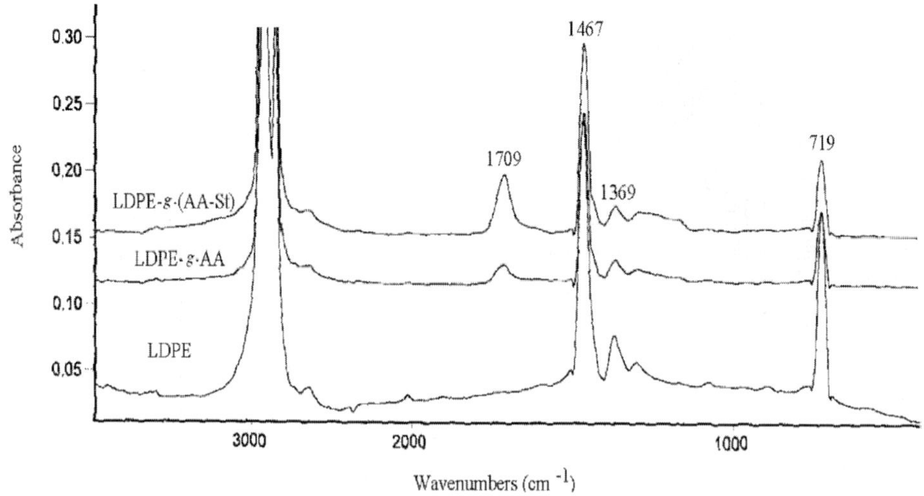

Figure 1.
FTIR spectra of LDPE, LDPE-*g*-AA and LDPE-*g*-(AA-St).

samples was much stronger than that for LDPE-g-AA, although the molar ratio of St to AA was equal to 0.1. The spectra showed that the addition of St could significantly increase the grafting degree of AA.

Effect of St/AA on the Gel Content and MFR of LDPE-g-(AA-St)

The effect of St/AA on the gel content and MFR of LDPE-g-(AA-St) is shown in Figure 2. The AA and DCP concentration were fixed at 5.0 wt% and 0.1 wt% based on LDPE, respectively. The MFR increased with increasing St/AA ratio and reached a maximum when St/AA was approximately equal to 1:1. This was because the addition of St could effectively decrease crosslinking reactions and prolong the life of the macroradicals. MFR changed only slightly when the St/AA ratio was higher than 1 because formed St radicals were enough to stabilize the produced macroradicals when St/AA was bigger than 1. As shown in Figure 2, the gel content is higher than 30 wt% in the absence of St although the initiator concentration is comparatively lower. This was because AA can easily auto-polymerize at high temperatures and auto-polymerization of AA decreases the consumption of DCP; consequently DCP would produce a greater amount of primary radicals. According to the grafting mechanism,[11] an increase in the amount of macroradicals leads to severe crosslinking of LDPE. Entanglements between poly (acrylic acid) and LDPE backbones also become severe. However, the gel content decreased gradually when the St concentration increased, especially at high St concentrations. As was reported previously,[13] St can easily react with the primary radicals produced by the initiator, stabilizing the radicals on the LDPE backbone and hindering the crosslinking reactions.

As shown in Figure 3, the gel content of LDPE-g-(AA-St) (St/AA = 1) is much lower than that of LDPE-g-(AA-St) (St/AA = 1/2). This is in agreement with the results shown in Figure 2, although there was difference in AA content in the grafting systems. Apparently, the addition of St can reduce the gel content of the LDPE-g-(AA-St) when the St/AA is equal to 1.

The gel content in the lower curve of Figure 3 is almost independent of the DCP concentration, when the DCP concentration is greater than 0.10 wt% because the grafting reactions under these conditions are dominating in the melt system. As shown in the upper curve of Figure 3, the

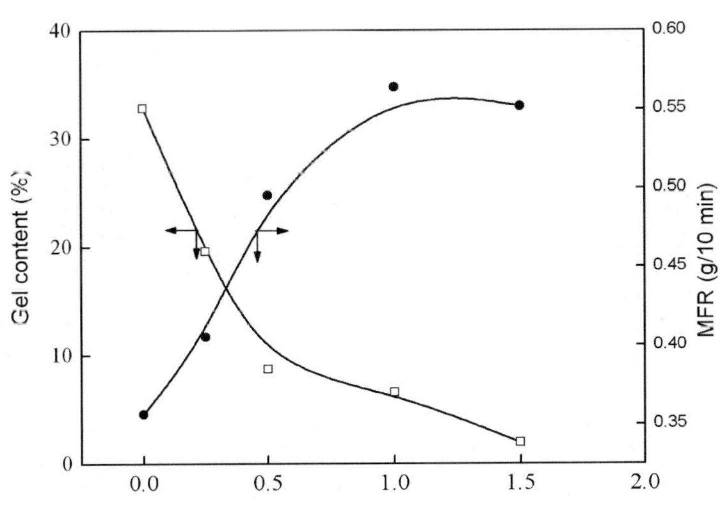

Figure 2.
Effect of the St concentration on the gel content and MFR value of the LDPE-g-(AA-St), AA: 5 wt%, DCP: 0.1 wt%.

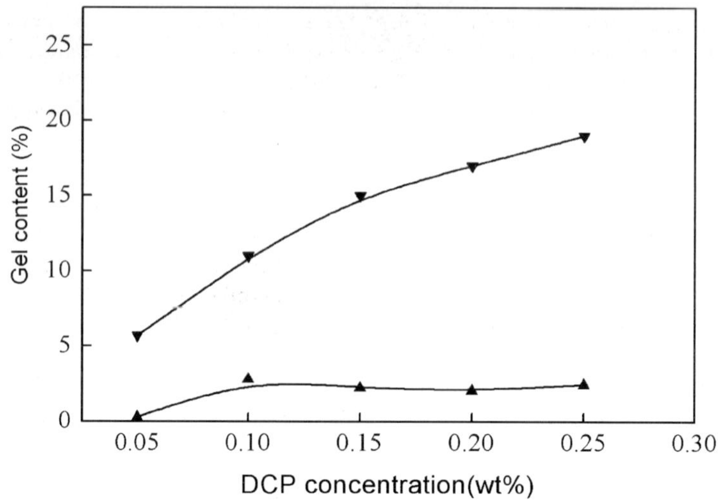

Figure 3.
Effect of the feed molar composition on the gel content, ▼: St/AA = 0.5; ▲: St/AA = 1, AA: 6 wt%.

gel content increases with increasing DCP concentration. This was expected when the DCP concentration was higher.

Effect of DCP Concentration on MFR and R_a of LDPE-*g*-(AA-St)

Figure 4 shows the effect of the DCP concentration on the MFR and R_a under equimolar feed composition. Increasing the concentration of the initiator leads to more crosslinking reactions and the MFR of LDPE-*g*-(AA-St) tails off. When the DCP concentration was about 0.3 wt%, the value of R_a reaches a maximum of 0.62. But the severe crosslinking reaction decreases the MFR. Intuitively, this was what should be expected as mass transfer became limited by the viscosity increase at high peroxide concentrations and grafting reactions gave way to crosslinking and

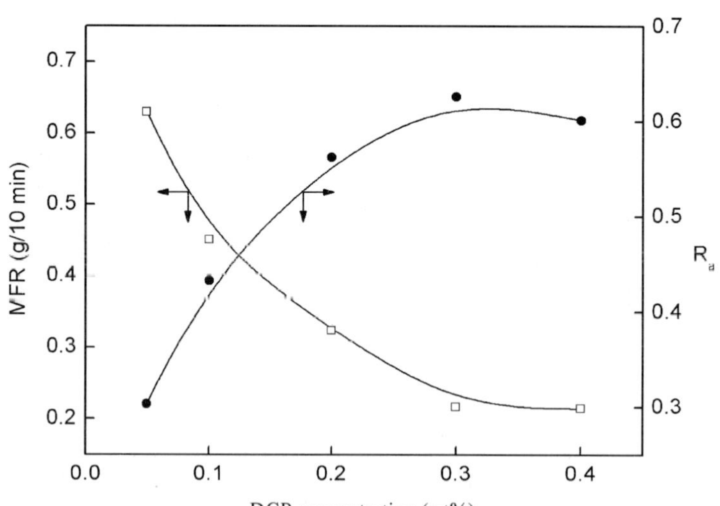

Figure 4.
Effect of DCP concentration on the MFR and R_a of LDPE-*g*-(AA-St), AA: 6 wt%.

gel-formation reactions. This effect led to an even lower value for R_a. Considering R_a and MFR of LDPE-*g*-(AA-St), a suitable DCP concentration was 0.2 wt%.

Effect of AA Concentration on MFR and R_a of LDPE-*g*-(AA-St)

Figure 5 shows the MFR and R_a of LDPE-*g*-(AA-St) versus comonomer (AA) concentration (St/AA = 1). The MFR of LDPE-*g*-(AA-St) was the lowest when the AA concentration was 1.0 wt%. This was due to the severe crosslinking reactions for the higher amount of DCP. MFR reaches a maximum value when the AA concentration is equal to 2.0 wt% and then decreases gradually. A reasonable explanation for this behavior is that the gel content decreases under these conditions. On the other hand, R_a was still smaller. When the AA concentration increases, crosslinking reactions become more important than grafting reaction. As a result, higher R_a values correlate with lower MFRs. Therefore, R_a initially increases and reaches a maximum when the AA concentration is about 5.0 wt%. At higher AA concentrations, the value of R_a starts decreasing slightly. This behavior is probably a result of the incompatibility between LDPE and AA. The polar AA monomer molecules tend to form aggregates dispersed in the LDPE matrix.[14] Furthermore, the AA would tend to homopolymerize noticeably. So the optimum AA concentration is about 5.0 wt%.

Rheographs of LDPE-*g*-(AA-St)

Torque rheometry has frequently been used to monitor chemical reactions during reactive melt mixing. The torque-time behavior of LDPE-*g*-(AA-St) (St/AA = 1) were measured to further investigate the effect of DCP concentration (from 0.05 wt% to 0.3 wt%) on the equilibrium torque. From the rheographs in Figure 6, we notice that when the DCP concentration is below 0.1 wt%, the maximum torque peak is absent. Increasing the DCP concentration caused the torque peak of the reaction to rise dramatically. When the DCP concentration was 0.2 wt% (Figure 6, curve c), not only the torque peak of the reaction increased dramatically, but also the values of the equilibrium torque were almost equal to that of curve b in Figure 6. This indicates that under this condition, the grafting reaction occurred to a suitable degree and that the crosslinking reactions were limited to a minimum degree.[16] However, when the 0.3 wt% DCP was added to the system, the crosslinking

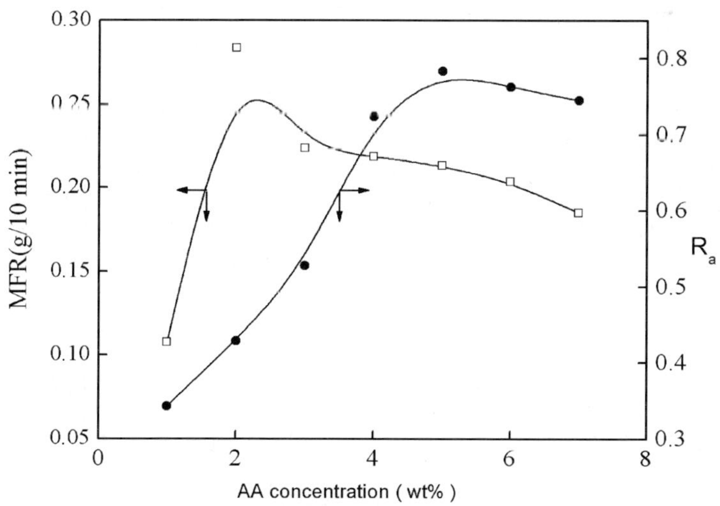

Figure 5.
Effect on monomer concentration on MFR and R_a of LDPE-*g*-(AA-St), DCP: 0.2 wt%.

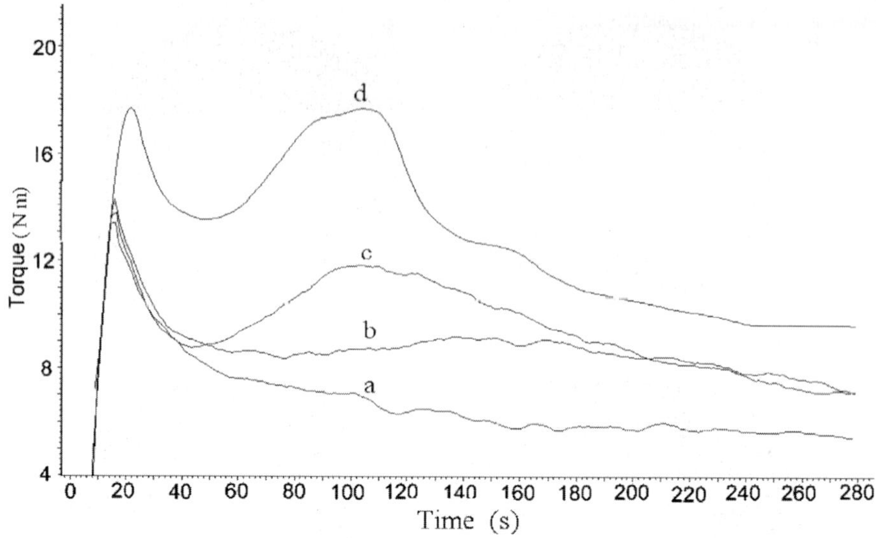

Figure 6.
Rheographs of LDPE as a function of DCP concentration, a: 0.05%; b: 0.1%; c: 0.2%; d: 0.3% DCP based on LDPE, AA: 6 wt%.

became severe for higher values of equilibrium torque. This is not desirable for processing with twin-screw or single-screw extruders.

It was observed from the rheographs that the torque is stabilized after approximately five minutes of mixing, suggesting that mixing and grafting reactions had occurred within these five minutes. In all, the suitable value of the DCP was about 0.2 wt% from both Figure 4 and Figure 6.

Hydrophilicity Characterization of the Grafted Films

The contact angles of water on the films of LDPE-*g*-AA and various LDPE-*g*-(AA-St) are presented in Table 1. The data shows that all the equilibrium contact angles of the samples tend to decrease when compared with 121°, the water contact angle of the control LDPE film. The contact angle of LDPE-*g*-(AA-St) is the smallest when St/AA is equal to 1:10. However, when St/AA is equal to 1:1, the contact angle increases again. When small amounts of St are added to the grafting system, there is an obvious increase of R_a of AA.[15] On the other hand, the St groups located on the surface of the film are comparatively lower. Both factors improve the surface polarity of the film and reduce the contact angle. When the St/AA ratio increases, the amount of St introduced onto the LDPE film increases and the hydrophobic benzene ring reduces the surface polarity of the film. However, as we can see from the Table 1, when the St/AA ratio is higher than 1:1, the grafting degree of AA is elevated concomitantly, but the St groups located on the surface severely hinder the hydrophibility of the grafted film. Therefore, the contact angle began to increase again. These results can be seen clearly in Figure 7.

Table 1.
Relationship between the contact angle of water, R_a and the feed molar composition.

St/AA	θ_e (Degree ± 2)	R_a
LDPE	121	0
0/1	108	0.2650
0.1/1	82	0.3843
0.5/1	90	0.4502
1/1	94	0.5705
1.5/1	101	0.5568

Rheological Properties

Capillary rheometry was used to characterize the control LDPE (St/AA = 0/0) and the

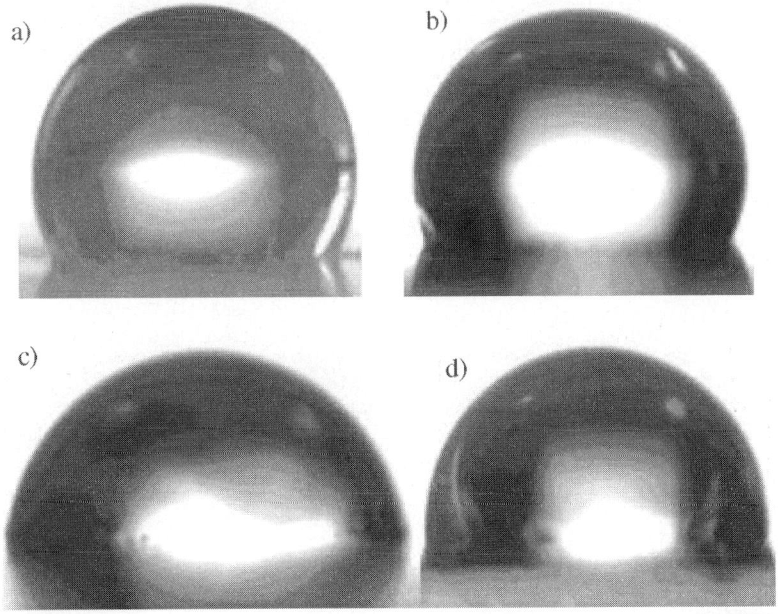

Figure 7.
Water drops on the films of LDPE, LDPE-*g*-AA, and LDPE-*g*-(AA-St), a: LDPE; b: LDPE-*g*-AA; c: LDPE-*g*-(AA-St)(1/0.1); d: LDPE-*g*-(AA-St)(1/1).

grafted LPDE samples. The viscosity curves as a function of the shear rate at 190 °C of the control and grafted LLDPEs with different St/AA ratios are presented in Figure 8. The viscosities of the control LDPE and its grafted products decrease as the shear rate increases, indicating a pseudoplastic behavior. LDPE-*g*-(AA-St)

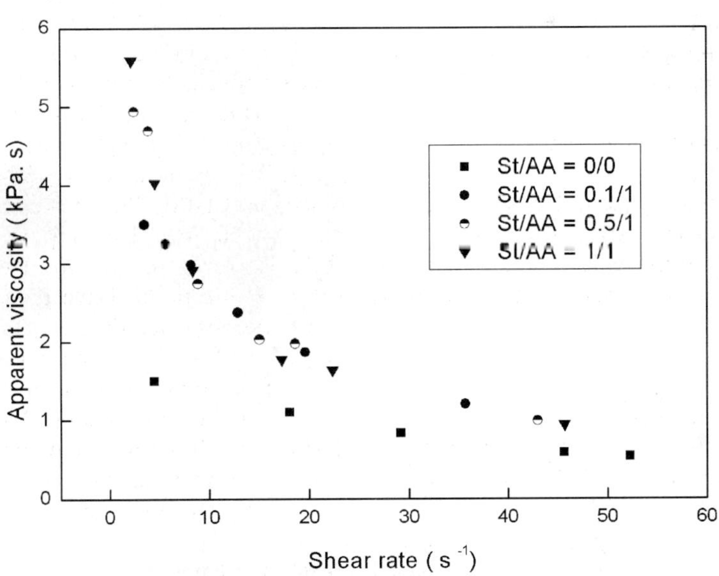

Figure 8.
Apparent viscosity as a function of shear rate for control LDPE and three functionalized LDPE with AA and St.

displayed higher melt viscosities with increasing St/AA ratios and a more pesudoplastic type behavior than the control LDPE, in particular in the low shear rate range which is more sensitive to interchain interactions. Similar effects have been found for GMA-grafted polyolefins[17] and this could be mostly ascribed to polar interactions between polar groups of AA in the polymers chains. However, the marked increase observed in the melt viscosity of LDPE-*g*-(AA-St) could not be accounted only by the effect of the grafting degree of the LDPE and suggested the possible occurrence of crosslinking reactions between the grafted chains in the melt.[18]

Conclusions

AA has been successfully grafted onto the LDPE backbone in the presence of DCP. LDPE-*g*-(AA-St) with high grafting degree of AA could be obtained when styrene was added to the melt grafting system. When the St/AA ratio was less than or equal to 1, increasing the concentrations of St improves the grafting degree of AA and the optimum molar ratio was 1. The flow properties of LDPE-*g*-(AA-St) could be adjusted by tailoring the St/AA ratio under this condition. But the hydrophobic nature of LDPE-*g*-(AA-St) was not consistent with the grafting degree of AA when St was used as a comonomer. When the St/AA ratio was 0.1, the LDPE-g-(AA-St) possessed better hydrophilicity. Suitable DCP and AA concentrations were about 0.2 wt% and 5 wt%, respectively for comprehensive properties of modified LDPE. The suitable modification time was about five minutes. Rheological properties indicated that LDPE-*g*-(AA-St) made at various feed compositions displayed higher melt viscosities with increasing St/AA ratios and more pesudoplastic behavior than the control LDPE samples, in particular in the low shear rate range. This copolymer with high reactivity and better flow properties could be widely used as a compatibilizer for LDPE and other polar polymers.

[1] J. H. Yang, Z. H. Yao, D. N. Shi, *J. Appl. Polym. Sci.* **2001**, *79*, 535.
[2] Q. Shi, L. C. Zhu, C. L. Cai, *Polymer* **2006**, *47*, 1979.
[3] M. F. Lu, S. J. Zhang, D. S. Yu, *J. Appl. Polym. Sci.* **2004**, *93*, 412.
[4] P. Mariano, C. Donatella, *Macromol. Symp.* **2004**, *218*, 173.
[5] Y. H. Wang, F. B. Chen, K. C. Wu, *J. Appl. Polym. Sci.* **2004**, *93*, 100.
[6] F. Pazzagli, M. Pracella, *Macromol. Symp.* **2000**, *149*, 225.
[7] X. Y. Qiao, Y. Zhang, Y. X. Zhang, *J. Appl. Polym. Sci.* **2004**, *91*, 2320.
[8] P. Ghosh, B. Chattopadhyay, A. K. Sen, *Polymer* **1998**, *39*, 193.
[9] H. L. Huang, Z. H. Yao, J. H. Yang, *J. Appl. Polym. Sci.* **2001**, *80*, 2538.
[10] Z. H. Yao, Y. Gao, J. H. Yin, *J. Appl. Polym. Sci.* **2002**, *86*, 2626.
[11] M. Z. Liu, Z. M. Liu, S. L. Ding, *J. Appl. Polym. Sci.* **2003**, *90*, 3299.
[12] M. W. Pan, L. C. Zhang, J. F. Yuan, *J. Appl. Polym. Sci.* **2005**, *95*, 419.
[13] Y. Li, X. M. Xie, B. H. Guo, *Polymer* **2001**, *42*, 3419.
[14] T. H. Kim, H. K. Kim, D. R. OH, *J. Appl. Polym. Sci.* **2000**, *77*, 2968.
[15] S. J. Kim, B. S. Shin, J. L. Hong, *Polymer* **2001**, *42*, 4073.
[16] H. L. Wang, H. R. Brown, *J. Polym. Sci. Pol. Chem.* **2004**, *42*, 263.
[17] M. Pracella, D. Chionna, *Macromol. Symp.* **2003**, *198*, 161.
[18] B. Wong, W. E. Bake, *Polymer* **1997**, *38*, 2781.

